I0406573

Laypersons Guide To Medicine and Formulas

Learn how to do your own formulas with the simple form of correlation and with the help & Faith of the Magi – ' Astrologer Priests & 3 Wise Men '

By Di Smith

© Copyrighted, 2007-2016, Sydney
NSW Aust.
By Diane P.M. Smith
dismith5002@gmail.com

Future Works in the Making
Analysing Nuclear Meltdowns by Astrology
Analysing Murder by Astrology
Analysing United Nations by Astrology

Contents

Authors Natal Chart

CHART 1

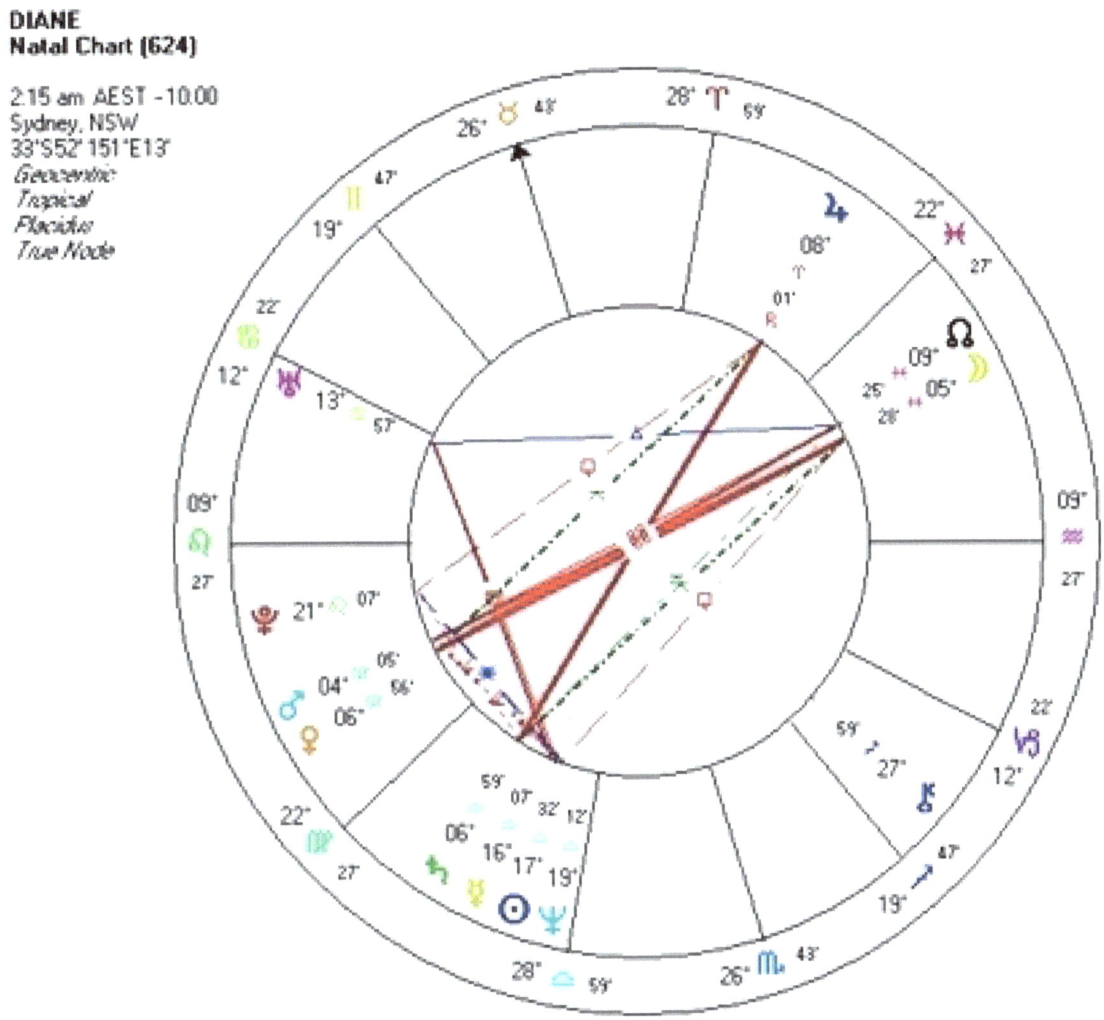

Australia's Constitution for NSW Natal & Transit Chart Given as Reference to- - ABS - Highest Causes of Death

CHART 2

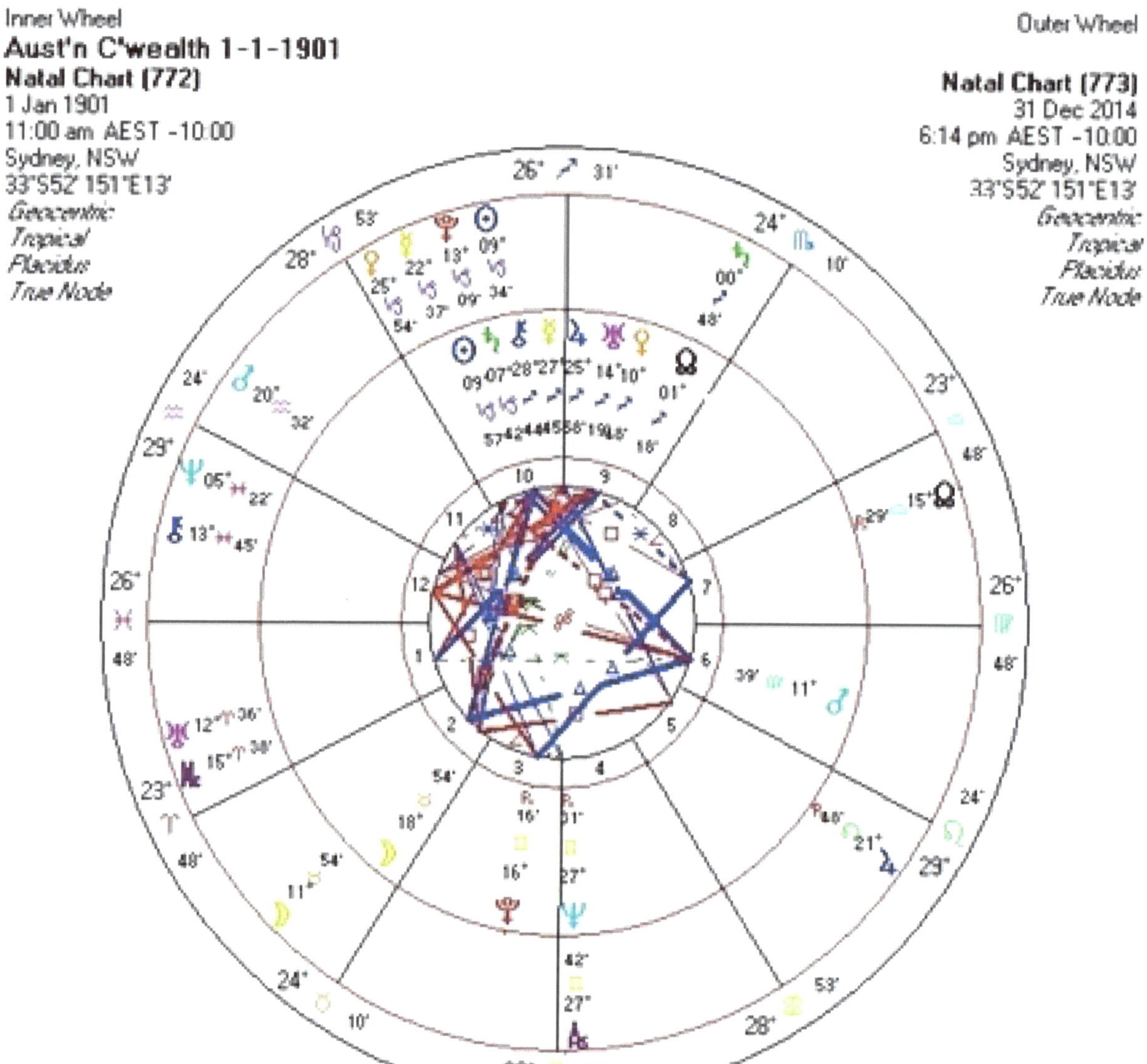

Australia's Constitution for NSW Natal & Progressed Chart Given as Reference to- - ABS - Highest Causes of Death

CHART 3

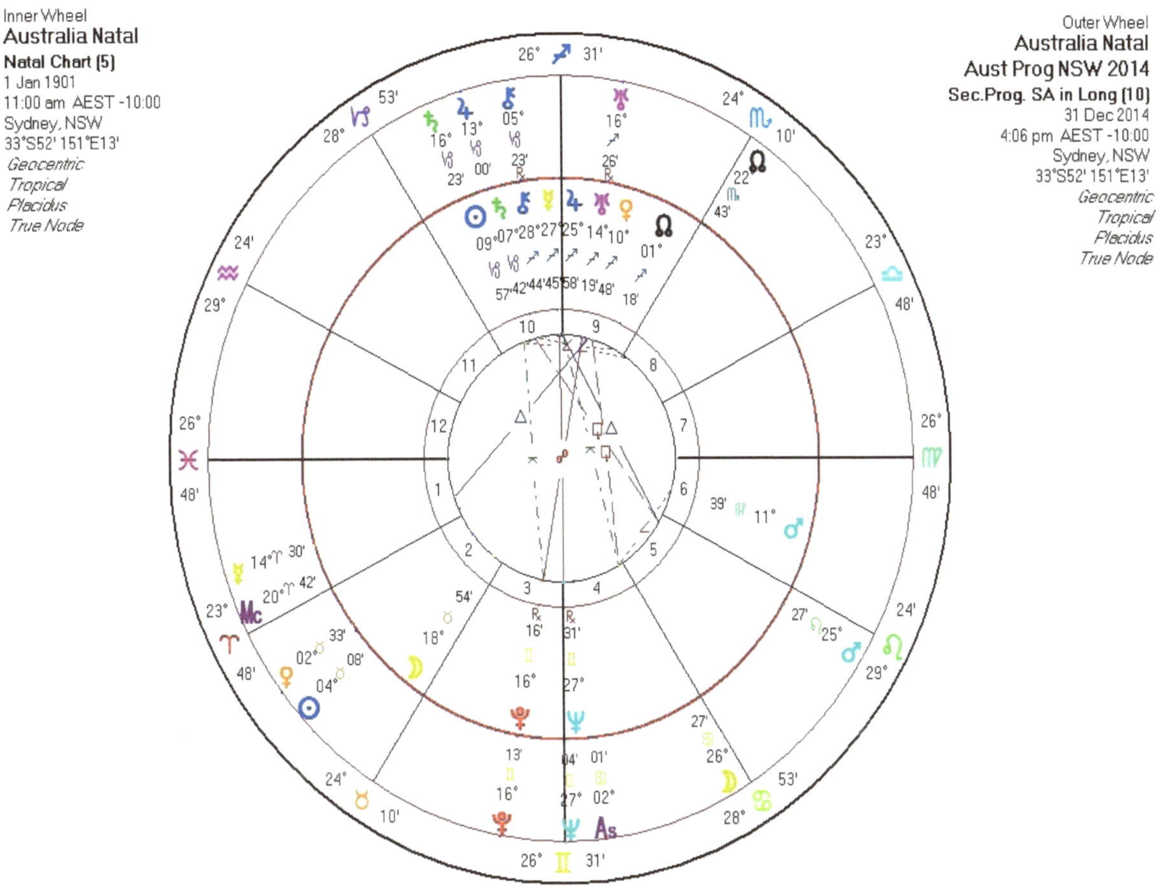

Australia's Natal/Progressed & Transit Flow Chart for 2014 this Illustrates the Flow of Planets

Natal/Progress & Transit	Degree & Minute	Constellation	Planet	House	Aspect
*T	00 48	Sagittarius	Saturn	9	⚻ Saturn - Venus
N	01 18	Sagittarius	North Node	9	⚻ Nth Node - Venus
P	02 33	Taurus	Venus	2	
P	04 08	Taurus	Sun	2	⚻ Nth Node - Sun
*T	05 22	Pisces	Neptune	12	□ Nth Node - Neptune
P	05 23	Capricorn	Chiron	10	
N	07 42	Capricorn	Saturn	10	
*T	09 34	Capricorn	Sun	10	□ Sun/Sun - Uranus
N	09 57	Capricorn	Sun	10	
N	10 48	Sagittarius	Venus	9	□ Venus - Mars
N	11 39	Virgo	Mars	6	□ Mars - Uranus
*T	11 54	Taurus	Moon	2	
*T	12 36	Aries	Uranus	1	
P	13 00	Capricorn	Jupiter	10	□ Jupiter - Mercury
*T	13 09	Capricorn	Pluto	10	
*T	13 45	Pisces	Chiron	12	
N	14 26 ℞	Sagittarius	Uranus	9	☍ Uranus - Pluto
P	14 30	Aries	Mercury	1	□ Mercury - Saturn
*T	15 29 ℞	Libra	North Node	7	
P	16 13	Gemini	Pluto	3	⚻ Pluto - Saturn
P	16 23	Capricorn	Saturn	10	
P	16 26 ℞	Sagittarius	Uranus	9	⚻ Uranus - Moon
N	18 54	Taurus	Moon	2	□ Moon - Mars
*T	20 32	Aquarius	Mars	11	□ Mars - Jupiter
*T	21 48 ℞	Leo	Jupiter	5	⚻ Jupiter - Mercury
*T	22 37	Capricorn	Mercury	10	
P	25 27	Leo	Mars	5	
*T	25 54	Capricorn	Venus	10	
P	25 58	Sagittarius	Jupiter	9	☍ Jupiter - Neptune
P	27 04	Gemini	Neptune	4	☍ Neptune - Mercury
N	27 45	Sagittarius	Mercury	10	/Saturn
N	28 44	Sagittarius	Chiron	10	

United States Declaration of Independence Natal and Progressed Flow Chart for 2016

America's Natal & Progressed Flow Chart for 2016

	Degree ' & Constellation	Planet	Hse	Aspect	
N	00 Cancer 31	Venus	8	⚻	Venus /Saturn
P	04 Aquarius 42	Saturn	3	⚻	Saturn/Jupiter
N	05 Cancer 27	Jupiter	8		
N	06 Leo 35	Nth Node	9	□	Nth Node/Mercury
P	07 Taurus 48	Mercury	6		
P	08 Gemini 22	Venus	8		
N	08 Gemini 49	Uranus	8		
N	11 Cancer 18	Sun	9	□	)
P	12 Cancer 31	Moon	9	□	) Sun-Mn-Ur /Nept
P	12 Cancer 38	Uranus	9	□	)
P	12 Libra 53	Neptune	11		
P	12 Aquarius 59	Pluto	3	⚻	Sun-Mn-Ur/Pluto Pluto in its Fall
N	14 Libra 45	Saturn	11		
P	18 Aquarius 39	Nth Node	4		
N	19 Gemini 56	Mars	8		
N	22 Virgo 23	Neptune	11		Nept in Detriment
N	25 Cancer 02	Mercury	9	□	Mercury/Sun
P	25 Aries 39	Sun	6	□	Sun/Moon
N	26 Capricorn 42	Moon	3		Moon in Detriment
P	27 Aquarius 06	Mars	4		
N	27 Capricorn 36	Pluto	3		
P	29 Pisces 10	Jupiter	5		

N = Natal P = Progressed

- A planet in their fall loses its strength and influence
- A planet in their detriment is limited in its expression of its basic characteristic:

America's Progressed Chart with Transit for 11 September 2001 the day of the collapse of the Twin Towers

CHART 4

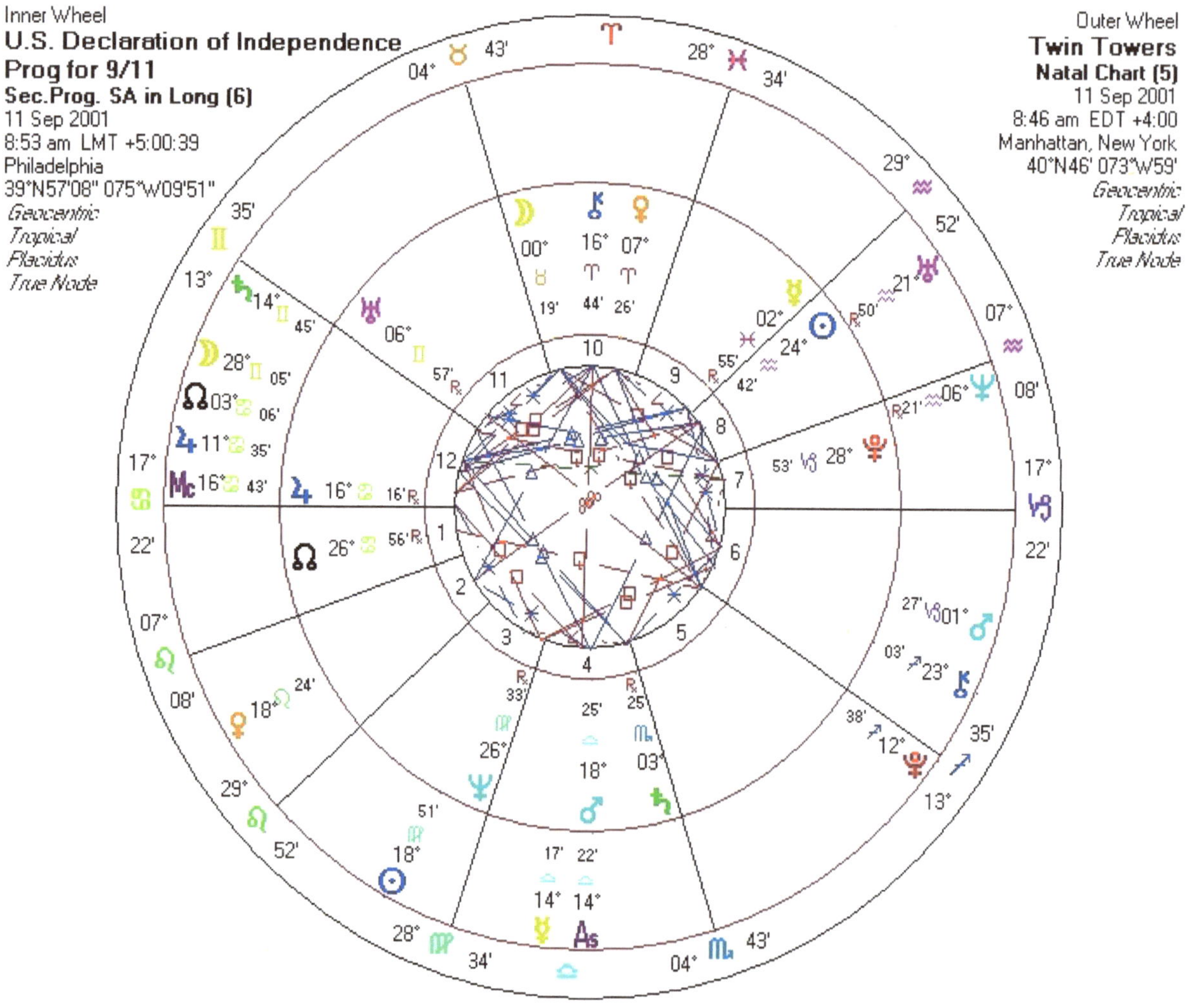

SYMBOL OF CONSTELLATIONS

Ram - Male Sheep

Aries

Scales + Male/Female

Libra

Bull

Taurus

Water Scorpion

Scorpio

Twins - Two Females

Gemini

Centaur Horse Man

Sagittarius

Crab

Cancer

Goat

Capricorn

Lion

Leo

Water Bearer

Aquarius

Virgin

Virgo

Fish

Pisces

SYMBOL OF PLANETS

Symbol	Name	Abbr	Symbol	Name	Abbr
☽	Moon	Mon	♄	Saturn	Sn
⊕	Earth	Ear	♅, ⛢	Uranus	Ura
☉	Sun	Sun	♆	Neptune	Nep
↓	Vulcan	Vul	⯓, ♇	Pluto	Plu
☿	Mercury	Mer	☊	North Node	Nod
♀	Venus	Ven	☋	South Node	Sno
♂	Mars	Mar	⚷	Chiron	Chi
♃	Jupiter	Jup			

List of Characteristics of Constellations from Aries to Virgo

♈♂ 1st	♉♀ 2nd	♊☿ 3rd	♋☽ 4th	♌☉ 5th	♍☿ 6th
I am	**I have**	**I Think**	**I Feel**	**I Will**	**I Analyse/**
Active	Artistic	Air	Babies	Bright/	**I Serve**
Active	Agriculture	Arms	Calcium	Shining	Analyse
Setting	Beauty	Brains	Crab	Confident	Colon
Courage	Earth	Breath	Children	Entertainer	Correctional
Direct	Farm	Communi-	Emotional	Entertainment	Critical of
Fire/Fiery	Land	cation	Emotions	Fire	Oneself
Head	Homie	Lungs	Family	Heart	Digestive
Impulsive	Jealous	Mimics	Homeliness	Humorous	System
Individual	Material	Monkey	Homemade	Husband	Earth
Action	Objects	Neighbours	Lakes	King/	Health
Initiative	Neck	/hood	Lime	Queen	Hospitals
Muscles	Land	Nervous	Mammary	Lazy	Judge
Own-self	Own	Nervous	Glands	Magnesium	Judg-
Oneness	Money	System	Memories	Proud	mental
Penis	Possessive	Paper-	Moody	Regal	Large -
Potassium	Practical	Work	Mother	Royal	Intestines
Quick	Possess-	Ribs	Mud	Self -	Menial
Sex	ions	Siblings -	Past	Control	Work
Scars	Same	Brothers	Rivers	Sensitive	Nagger
Sheep	sex	/Sisters	Sensitive	Solar	Practical
Work place	groups	Theft	Subconscious	Plexus	Pure
using sharp	Sensual	Transport	Stomach	Son	Snakes
instruments	Stubborn	Twins	Tenacious	Stubborn	Virginal
Wounds	Throat	Versatile/	Water	Suns	Work
made	Thyroid	Versatility	Womb	Rays	Work
By sharp				Teenagers	Place
instruments				Warm Hearted	

List of Characteristics of Constellations from Libra - Pisces

*

♎/♀ 7th	♏/♇ 8th	♐/♃ 9th	♑/♄ 10th	♒/♅ 11th	♓/♆ 12th
I Balance	**I Create**	**I Perceive**	**I Use**	**I Know**	**I Believe**
Air	Bowel	Archer	Ambitious	Airplanes	Alcohol
Arts	Brave	Exaggeration	Business	Air	Cold
Balance	Cruelty	Expansive	Bones	Ankles	Confused
Beauty	Envy	Fire	Carbon	Circulation	Drugs
Borderline	Funerals	Foreignors	Contracts	Fixed	Fear
Borders	Graves	Freedom	Control	Electrical	Feet
Contracts	Jealousy	Gestures	Cold	Eruptions	Film
Cooperative	Masses	Hair	Dominant	Freedom	Fish
Courts	Others -	Hips	Earth	Gambler	Ghosts
Diplomat	Money	Hopeful	Efficient	Groups	Guilt
Friends	Possessive	Impatient	Goal -	Humanitarian	Hotels
Helpful	Priest	Laws	orientated	Intellectual	Inferiority
Indecisive	Priestly	Legal	Goats	Open Minded	Instincts
Kidneys	Pope	Liver	Honors	Progressive	Institutes
Lawyer	Rape	Outdoors	Knees	Radiation	Lakes
Marriage	Real Estate	Outgoing	Marrow	Risk Taker	Losing thngs
Mediator	Reproductive -	Philosophers	Objective	Salt	Lymph -
Music	Organ	Philosophy	Orderly	Shin	Glands
Ovaries	Regeneration	Politics -	Orders	Stubborn	Missing
Partnership	Reserve	from	Professional	Sudden	Pictures
Peacemaker	Resourceful	Overseas	Reliable	TV	Photographer
Pleasure	Water	Punter	Ribs	Uranium	Sleep
Police Officer	Sarcasm	Religion	Self Centred	Uncompromis-	Spiritual
Refinement	Scorpion - Water	Silica	Serious	ing	Sponge
Relationships	Spite	Sports	Solitary	Unpredictable	Subconscious
Scales	Stubborn	Sports Arena	Structure		Surrender
Soda	Sulfur	Theatrics	Vertebrae	Volcano	Swimming
Sweetness	Transformation	Thighs		Water -	Tired
Urinary -	True Friend	Travel		carrier	Unconscious
Track	Truth	Undiplomastic		Wilderness	Water
		University		Windy	

Chapter 1 - Australia's & Ors Health Aspects for Sceptics

For people who wish to be healers it will take a leap of faith to make formulas by using Astrology, let alone to use the Runes as I've done to create this book and to confirm the formulas I've made, which I've detailed under Chapter 12 - 'Explanation to Formulas and Examples...' but the simple formulas below should change your mind if you ever questioned 'if God was so great, why couldn't he heal everyone?' but if you try the Zodiac symbols you'll find he/she has, you should get immediate relief, of course there needs to be the problem.

In saying I've used the Runes not everyone can use them, my planets show I'm drawn to anything spiritual, whilst I have good and bad influences to Astrology, with a Uranus square to Mercury /Sun/Neptune means I get distracted easily especially by the TV or groups in the family or where I'm living. It means if I don't put a conservative effort into something my brain won't work, and once I get an idea I must write it down because my thoughts change so rapidly. If I was doing something that needed no thought I'd be okay, but probably even packing boxes I'd stub a finger, I must admit I've been obsessive compulsive, in saying that I will pick up a Rune to confirm I'm going into the right area, just like a doctor might pick up a medical book before picking up a scalpel. I feel that the better part of my life my main aim has been to show Astrology alone can stand on its own laurels and be a scientific tool to helping humanity and to prove the scientific world wrong when they say it's a pseudo-science, when they say this basically they are saying our values have no worth as the philosophies are the teaching we live by.

I was given a renewed enthusiasm when I read India's government wanted a rewritten proposal for Astrology to be accepted as a creditable source or reference. In the western world the governments acceptance would enable it to be taught in schools and universities and there'd be no end to its discoveries in health, environment, planetary upheavals, and discovering how to solve countries wars just to name a few.

Personally I've found some Astrologers appear arrogant which I put down as a shield that is put up to protect them; some Astrologers have exquisite knowledge and own the arrogance. One of the things I do believe is that Astrology does show us how we affect our solar system if we study the minerals, it shows how one planets action will affect another out in the atmosphere in a physical sense, so it's a useful science for Astronomers / NASA and World Health Organizations, i.e. currently Pluto/Plutonium will have a direct affect on the Moon. With Pluto being in Capricorn opposite Cancer we can expect a breaking down of the Moon's surface as the Moon rules Cancer causing it to possibly go powdery, whilst I haven't studied what is going on with the Moon at the moment Pluto's influence is the breaking down of earth substance Capricorn an earth sign it's a ruler of carbons, which is being revolutionized by the Pluto influence. But, in speaking of the Moon it could be more affected when countries are doing Nuclear and Plutonium tests more than any other time. I know black holes have been noted on the Sun and it's believed to be a loss of temperature, I know that Neptunium having been named after Neptune is being used in spaceships and from an Astrologers perspective believe it would cause the Sun to loose heat and would be a health aspect to the Sun. In our stars Pisces the constellation Neptune rules is 150° from Leo the constellation that rules the heart and so Planets that are in this range should be paid special attention to.

For me this book it a testimony to Astrology and gives a practical outlook in finding formulas, you will notice I have broken down each section of Astrology in areas people wouldn't have even realized it covered, to show you how to combine them, each having a planet/constellation/house specific to them, it will show you why maybe tomato sauce could be used, where tomatoes have over 20 reported medical uses, and tomato sauce is stirred and boiled in water requiring cardinal/action-heat and water elements. These are the elements I use in preparing formulas. It is in some way far distant from being a pharmacist you could expect, whilst I suspect a pharmacist would understand with a mandate to study biology and chemistry, of course in some respect they are far superior as they would even be able to give the chemical breakdown. Whilst if pharmacist were that good they'd be able to tell you every element causing an illness and the areas to look this is what Astrology does. And, by studying every aspect of it, it will show you the foods to use and how to prepare them.

For formulas I've found I have to combine colours and you will have to learn the colours. From the colours of Constellations/Planets & Houses you need to find the foods under those signs which combine and have the same colours of the Constellation/Planet or House. You may find you're looking for something which is air sign which could lead you to a gas, that's once you know what air signs are. In one formula I used the Blue Claw Crab, this gave the colors of both Cancer and Aquarius. Astrologers are half way there. Shortcuts can be made with Tissue/Cell Salts but they are banned in Australia, whilst you can buy the products separately, individual requirements would be specific. The three you can buy over the counter at a supermarket is table salt, silica and magnesium.

In the early 90's I researched Current Affairs and explained the issue using Astrology on radio, here I realized how important word association was. At the time a scientist discovered Mussels were good for Muscles (aka My·o - Muscle in medical terms e.g. Myocardial = muscle tissue in the heart) and I was able to relate this to Aries which rules Muscles in the anatomy, ever since I've used word association in my search for formulas. Having studied Physiology and Anatomy and following this theory I found medical terms often had two explanations one of course referring to the anatomy and the second going into Latin or Greek would give a reference to a plant or even animal. My increased interest was insatiable but internet explanations seemed to lack depth and my lack of ability to retain what I'd read was more annoying, whilst I wrote down what I'd found it wouldn't have made, if a page.

As for Sceptics they often need proof and I can't say I haven't been a sceptic myself, but, what I'm about to share with you is my greatest secret and the simplest of formulas, and I'd like to be given the credit for finding them after they had be written in black and white for thousands of years without us even realizing it.

Would you believe the Zodiac symbols of Constellations give us our first and most important clue, but there are exceptions, their cell salts are under Tissue/Cell Salts Combinations, the Zodiac Element will also give you a process required to creating a formula which is included in Chapter Two and Seven. Initially I'd found it hard to relate symbols that were human so I'd look to its elements and meaning i.e. Gemini is an air sign rules the hands, so I looked for something that grew in the air and Comfrey flowers were called 'hands' I found they substituted well. Then I realized Gemini also ruled the brain and lamb brains were easily bought – they also worked. I used the same technique for Virgo. The symbols are as follows, but, pain is ruled by Capricorn which is goat/oil - mixture recommended:

1. Aries symbol is the Male Sheep, a Ram - it rules the head and muscles; if you have temperature or a cut, problems with muscles, look to the Ram for a cure, alternate mussels for muscles.
2. Taurus symbol is the Bull, the Male Cow - it rules the neck and thyroid and growth hormones. The oil of a cooked piece of meat does wonders for sore throats and thyroid problems.
3. Gemini symbol is the twins. It has to do with humans - so naturally I've looked at alternatives which has led me to different areas. It rules lungs/Respiratory System/brain and arms. I've used sheep's brains, brains being the optic word; but also I found the Comfrey flowers were called a 'hand' the flowers grow in the air and will help with the brain; it's an air sign so oxygen as well.
4. Cancer symbol is the Crab - it rules the stomach/breasts/mammary glands and digestive system; on the psychological level memories & helps to control worry and the hydrochloric acids of the stomach. Because the stomach ferments I'd look to fermented Crab. And, suggest it could be trialed for Cancer. I did use it in a formula for the Person in a Coma.
5. Leo symbol is the Lion and China already has its cousin the Tiger medically recorded as being honored - in using colour and element association the Lion is safe, Leo's element is hot, it's Planet is the Sun it's colour is orange and I've experimented successfully with Oranges numerous times and found if you squeeze one in hot water it immediately removes symptoms of heart problems. There are various ways of using any of it because you can also put a slice of orange over the heart to strengthen it.
6. Virgo symbol is the Virgin it refers to purity - in the anatomy it rules the intestines, it's ruler is Mercury and it's an earth element. In the past I've used foods that are long and winding i.e. long beans. I've never tried it but I believe a minute dosage of Mercury could be used on the intestinal region. Also as it is an earth humate and ligmate. There is also Chitterlings made from pig's small intestines, some countries use the intestines to make these foods, for intestinal problems I'd suggest them.
7. Libra symbol is the Scales held by a person, they are metal it rules the Kidneys - where I've used Kidney Beans and Kidney meat easily obtained from the supermarket. But something simpler is Soda aka Natrium (Nat).

8. Scorpio symbol is the Water Scorpio, I have yet to find a water scorpion but I have seen them in documentaries as far as I know so far they are not being used for the reproductive system, but would recommend it. As it is a water sign ruling the masses I'd look for something that are in huge numbers that are fished and either very small or large. It also rules the underworld and poisons where you could look at green potato peel which grows under the soil.
9. Sagittarius is the Centaurian, the Archer half man half horse. As Sagittarius rules the liver and skin/hair and nails along with Capricorn; it stands to reason the horse's Liver or its bile would be found to alleviate liver problems, this would be the best but, I usually just buy the cows Liver from the supermarket or Chicken Livers.
10. Capricorn symbol is the Goat, the part of the body it rules are the knees/bones and marrow. Using the marrow from the Goat especially helpful to rub into the vertebrae and knees. Also you can use the oil from Goat meat along with eating it.
11. Aquarius symbol Water Bearer. Aquarius rules the circulation and ankles. I've found its chemical association is Nat Mur, aka Sodium Chloride, aka Table Salt as such for circulation table Salt is required.
12. Pisces symbol is the Fish it rules the lymph system and feet and negative symptom is confusion, so obviously I've used Fish for problems with the white cells and lymph.

In outlining the areas that can be helped by the zodiac symbols I have predominately made oils that can be rubbed on specific areas. I'd like to move to an area which involves charts. Because these are what formulas are based on. And, it wouldn't be enough to just say something for a sceptic to believe it, so I'm moving onto Australia's Health issues, the following is from the ABS for various years, it's a record of what Doctors have recorded as leading diseases to sudden death and Chart 2 in the preliminary section shows Australia's chart both Natal and Progressions, I further added a Flow chart.

Australia's Leading Causes of Death

We don't often realize that the area we live can be a major problem to our health. This part shows how a countries planets show a potential for illness, it will help you to understand the theory behind the formulas. It's the close orbit of the planets and their degrees which foretells which action will take place first.

In the table below it shows the leading causes of death in Australia to be Ischaemic Heart Disease which includes angina, blocked arteries of the heart and heart attacks. The death number for 2005 was 20,570 rose in 2009 to 22,587 and decreased in 2014 to 20,173, it's been said to be steadily increasing since. This is interesting because planets do move and a lot slower when looking at progressed influence or the outer planets, where I use the Sun's ☉ progressions for all planets. Planets also known to go retrograde - backwards but whether they are just going up and down at the same degree is an ongoing argument. As for progress instead of the Sun moving 1 degree per day which is the movement of the Earth around the Sun which the Natal Chart is based on, it moves 1 degree in a year, when I haven't got quick reference to a computer what I do is refer to my Ephemeris which gives the planets movements e.g. if a person is 45 and I want to see what their progressed chart looks like I'll count 45 days from the person's date of birth (D.O.B.) and draw a line under the day and planets that it refers to.

An Ephemeris is written by Astronomers the ones I have were written by Scientist at NASA space station. They give a 50-year analysis of where the planets are or will be, in fact I've seen one from 1900 to 2000. For further information on movements refer Chapter Six, 'PLANETS MOVEMENTS' which explains how fast the planets move around the Sun over a year this will help to understand why it appears a planet hasn't moved.

When analysing a problem there will be 3 charts I look to 1. the Natal based on time/date/place of birth of an accepted Constitution which shows complete change. 2. The Progressed and 3. The Transiting chart where planets are today or at the time of incident these are the 3 charts I refer to in Chapter Twelve. When doing a chart, I look at Natal with Transits or Transits with Progressed and Progressed with Natal. For this reason, you can get a triple affect in looking at a chart e.g. Saturn ♄ progress (outer wheel) is 9 degrees from Australia's natal Saturn both in Capricorn ♑, there will be this triple affect when transiting Saturn now in Sagittarius moves to Capricorn and will be conjunct natal Saturn on the 24 November 2018.

LEADING CAUSES OF DEATH – Australia – Selected years – 2005, 2009, 2014

Cause of death	**2005**	Rank	**2009**	Rank	2014	Rank
Ischaemic heart diseases	20 570	1	22 587	1	20 173	1
Dementia, incl. Alzheimers	4 653	5	8 280	3	11 965	2
Cerebrovascular diseases	11 513	2	11 216	2	10 765	3
Trachea, bronchus & lung cancer	7 399	3	7 786	4	8 251	4
Chronic lower respiratory diseases	5 428	4	5 984	5	7 810	5
Diabetes	3 529	8	4 176	6	4 348	6
Blood & Lymph cancer (including Leukaemia)	3 614	7	3 811	8	4 275	7
Colon, sigmoid, rectum & anus cancer	4 171	6	4 068	7	4 169	8

.

To create Australia's chart, I've put into my Astrology program, Solar Fire the believed time of our first Constitution as 11:00 am, based on when New South Wales accepted the Constitution on the 1 January 1901. The time can be argued but as Australia's Ascendant is in Aries which is the position the normal Constellations are placed, the problem should be minor. It's said 'if you don't know the time of birth start with Aries'.

The planets degrees, except for the Moon are still the same whether it's N.S.W or other States who've accepted it at different times only the house position will be different. In the above charts I've not only worked on natal chart/progressions but also transits for when ABS gave the last statistic in 2014.

Really the population affected by the diseases is a drop in the ocean because of our population, if you calculate the spread over the States and Territories, per year it has raised to 20,173 we have a population of approximately 36 million, according to www.worldpopulationstatistics.com/population-of-australia I might add this greatly differs from the ABS who recorded 10 million less. In saying this most countries have a great deal higher population and their illnesses are far greater, as for instance America's population is 320 million as compared to our 36 million but all this is subjective. Whilst you mightn't be a believer of Astrology the population difference can be explained by Pluto inconjunct Saturn, Pluto rules the masses and when Pluto inconjunct Saturn shows problems with calculations; calculation and clerical areas are covered by Gemini the constellation Pluto is in and there is Saturn in Capricorn both having the influence of constraint involving the government sector. The clerical department also wouldn't be given correct information from communities along the ocean coast Uranus – groups and huge numbers of people in Sagittarius – sandy areas opposite Pluto

In this section with the focus on Australia I'd like to show how we can see the illnesses by the different planets/constellations/houses and what makes things easy is that 'death' is ruled by Pluto. I'm not exaggerating, in mythology it's well known to be the ruler of death and the underworld, along with Scorpio and the 8th house, these are areas we all have in our charts. But, Pluto will be the focal point for looking at death in countries. In the external world it rules the greatest powers that be, this can be positive and negative.

<u>Heart Attack – Heart ☉ ♌ 5th House</u>

First the meaning of the Constellations/Planets/Houses have to be understood, a breakdown is given in Chapter 2 explaining rulership and further detail is throughout various parts of the book, this should be memorized, by following it we know in Astrology for heart attacks, Ischaemic diseases, we are looking at the Leo ♌ Constellation, Sun ☉ the planet, and the 5th house, which is the highest problem area in Australia, to see what is troubling that area I would then be thinking of the constellations which naturally oppose and square being Aquarius/Scorpio & Taurus and notice from the flow chart, progressed Mars in Leo/5th house, nothing too drastic with that, but it would mean in most cases the heart would be over heated, both elements give out heat Leo and the 5th house, which you can actually feel with your hand going over your heart, and this is intensified as there is a trine 60° angle to Jupiter another hot planet.

So now I look to the influences that will block, this can be gradual or sudden. The planets/constellation & house that build blockages and acts unpredictable are Saturn ♄/Capricorn ♑ & 10th house causes calcification/blockages, blockages cause pain and Uranus ♅/Aquarius & 11th house is freedom loving or unpredictable, and the other influence relative to death is Pluto, (please note - throughout the reading I later only refer to a planet or constellation, but I do mean the whole 3 influence if I'm writing generally, and whilst I look to obstruction each has a good side but we are looking at detrimental areas so it may seem more pointed) following where Saturn/Capricorn & 10th house is involved it's normal influence is to organize. But we also have Pluto which is fixed and powerful. For detrimental aspects causing health issues I look to planets and degrees that are opposite ☍, square □ and 150 degree angles with the symbol ⚻ and 210 degrees inclusive. This is often the surprising part that can help doctors.

Now let's look to Australia's Chart 2 and Flow Chart in the earlier part of the book. What I see first is that the Sun ☉ ruler of the heart is in Capricorn ♑ instigator of pain, where the planet conjuncting (same degree) is Saturn ♄ . To emphasis this, it is also the planet which rules Capricorn and it's seen in the 10th House wooow, the house is only correct if my time is correct, what this would mean is there's a lot of pressure to work, to build to have structure, there would also be a lot of constraint, this can be seen by the Police who act as a controlling force and as debt collectors. Further progressed Saturn is making a quincunx to Pluto ♇. This is shown by the influences of Pluto both in progressed and natal chart, outer and inner wheel of chart 2. Gemini ruler of thoughts generally plays on Scorpio whilst Capricorn – structure /control plays on Gemini as Gemini has to do with speech and freedom to express one's thoughts it can only find illness if constrained of the airways. When Saturn is in the earlier degree it means air would be blocked, and the hearts function would be blocked by Pluto.

<u>Reasons for Perspective Respiratory Problems - ☿ ♊ 3rd house</u>

The fact that Pluto is in Gemini would contribute to problems with breathing and adding Progressed Mercury is squaring Transiting Pluto would cause considerable constraint to people who are liable to fall ill to respiratory problems. Mercury ☿ being the ruler of Gemini ♊ gives a double influence where the anatomy covers Lungs/Brain and Arms. Transiting Pluto is in a sign for 13 years, so it will gradually move on. But there are also the oppositions we must consider when Venus and Uranus are both opposite in Sagittarius, Sagittarius is expansive and the lungs would find problems with Venus which rules the thyroid/the growth hormones and kidneys and Uranus which produces salt and contribute to unpredictable actions of the Liver. It also means that the neighbourhood is having problems expressing themselves especially where foreignors are concerned. Gemini and Mercury are both mutable signs which means to talk, Venus also means money where schools might be provided for but not physical neighbourhood which are being choked by the incessant building and constraints put on by government.

Over 115 years <u>Saturn ♄</u> has progressed back to where it had been conjunct the Capricorn Sun ☉. It's now at 16° making that quincunx angle to Pluto♇ **which is also 16°**; in Gemini ♊ something that is very important to notice sudden death as it is opposing and prior to Pluto is Uranus ♅ Uranus has an unexpected explosive nature, at least for some it's unexpected, the rumblings of a volcano gives forewarning the same for a person or a country, it would mean a blockage of the fluids brought be circulatory system would be stopped to the brain. These two planets are two fixed immovable impenetrable forces The 150° in itself gives an understanding to why Ischaemic disease is the highest death rate in Australia and why Respiratory problems have increased,

The Sun squaring progressed Mercury ☿ would show magnesium stopped by the airways and with Progressed Mercury being in Aries – Sun and Aries both Fires signs show a heating up of the muscles by phosphorous and magnesium, this would dry up the muscles of lungs and brain, all adding to respiratory problems and to the thinking process.

There is a close relationship between respiratory problems and dementia as the constellation square each other so each can be a cause of the others problem.

Dementia including Alzheimers - ♆ ♓ 12th House

Neptune ♆ ruler of white cells and lymph brings with it confusion when opposed, a negative effect when opposite Mercury, shows constraint in air, constraint in speech. Transiting Neptune ♆ is 11 degrees Pisces ♓ (I give a 10-degree orb this is exceptionally wide) and is squaring Transiting Saturn which is in Sagittarius, the transit is also squaring Natal and Progressed Pluto. Neptune is one of the out planets and is slow moving it will take some years before it will move passed Pluto, with Neptune ♆ being in Pisces ♓ the constellation it rules would cause communications to be more confused having this square to speech, and would give reason to why Dementia disease is a rising factor in Australia.

Because of the degree and house where Pluto is seen we can also surmise that air, the element of Gemini is being blocked by Pluto, the liver would be working unpredictably and it would have allowed calcification to build up. Uranus is only 3 minutes past Saturn at a 30° angle. Where the affect would be to make the liver work harder. Pluto opposite Sagittarius would stop the air from reaching the liver apart from it working unpredictably, whilst signs are opposite they are dependent on each other to run properly.

All may seem severe in itself but these influences are relative to a person's/countries chart and to me it gives an understanding to why Australia's no.1 sudden death killer is problems with the heart.

Further the additional influences of transits surrounding a chart can be seen with Americas chart it has considerable amount of problems as I will be going into, but, it's nothing if you compare the natal/progress chart with the transits i.e. those of the 9/11 Twin Towers and their unexpected collapsing caused by two hijacked airplanes, I won't be going into this but I have included their chart to show the caos.

America's Health Aspects

CHART 5

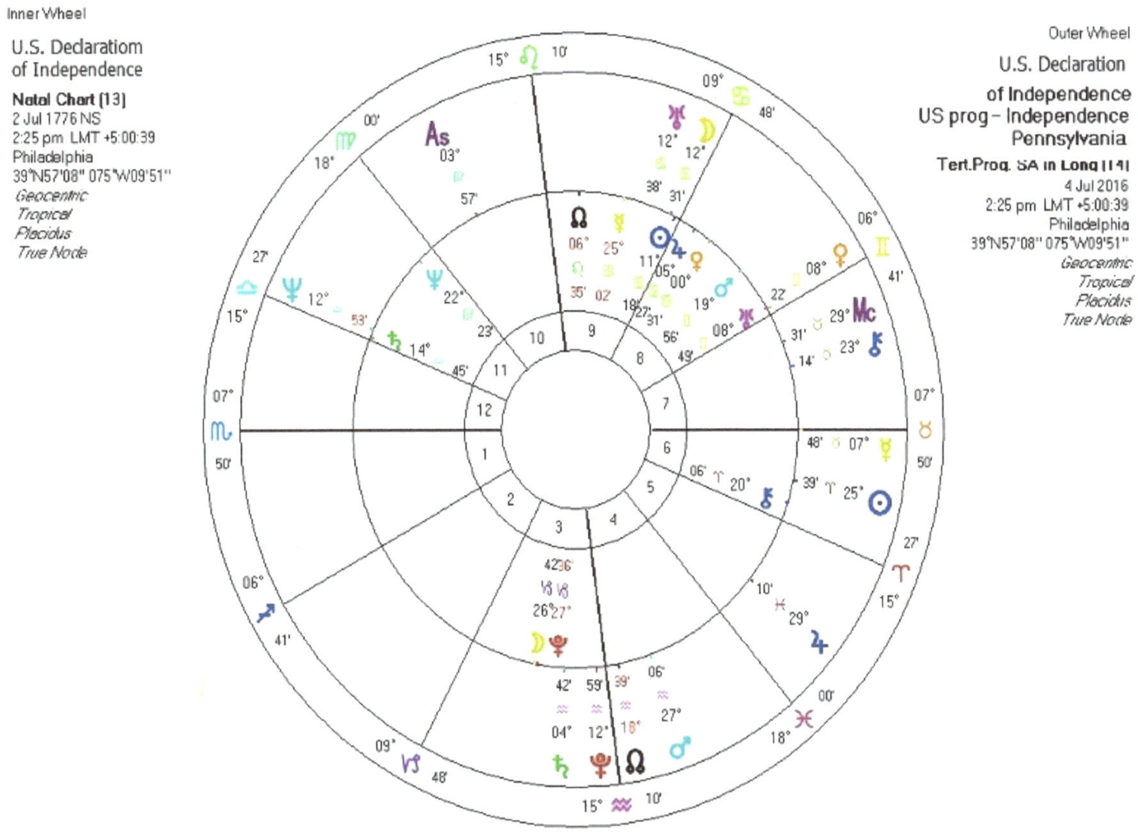

The chart is based on United States Declaration of Independence Natal Chart for 2 July 1776 – and Progressions for 4 July **2016.**

If this section seems a bit disjointed it's because I had no headings over each illness when I began and I'm going to work backwards to add the additional illnesses to Australia's problems; in some ways one illness will just flow onto another because the planets are close, but I realized each illness is defined by a planets, so please stay with me, if you look at my chart Uranus ♅ squares my Mercury ☿ /Sun ☉ and Neptune ♆ so I've had to work through a lot of personal disruptive influences just to get this book out.

America, going by Medical News Today http://www.medicalnewstoday.com/articles/282929.php records the highest natural cause of death is also Heart Disease the same as Australia's as seen below, this could be the result of both countries having an English Parliamentary system, a system which is still seen today with some changes, the range of diseases below.

Leading causes of death for males and females, 2016

	Deaths
Heart disease	611,105
Cancer (malignant neoplasms aka tumour) -	584,881
Chronic lower respiratory disease	149,205
Stroke (**cerebrovascular diseases**)	128,978
Alzheimer's disease	84,767
Diabetes (diabetes mellitus)	75,578
Influenza and **pneumonia**	56,979
Kidney disease (nephritis, nephrotic syndrome, and nephrosis)	47,112
Suicide (intentional self-harm).	41,149

When we know what the illness is we can then look to what is causing the problem. For instance, heart disease is generally caused by calcification in the arteries narrowing the blood supply, something I should have explained in Australia's chart, causing calcification are planets in Capricorn ♑, the thing is with astrology it doesn't change, e.g. if the problem is Thrombosis which is blood clotting in the circulatory system, we often find people having to wear tight stockings in the legs, the lower part of the leg is ruled by Aquarius, circulation is Aquarius so I'd be drawn to check that area first, not Leo for the heart, the problem with Thrombosis is the fear of the clot moving to the brain or lungs, which I believe would bring in Neptune – fear and Gemini/Mercury/3rd house the ruler of – lungs/brain and thoughts.

Heart Attack

In Americas flow chart I see Sun ☉/Moon ☽/Uranus ♅ conjunct, and quincunx progressed Pluto ♇ but, only 2 degrees from squaring natal Saturn ♄ another cause of calcification and constraint. This is just after Pluto in its Fall, which would cause its energy to be more negative, seen in the Flow Chart at the front of the book, page v. The natal Sun ☉ having progressed Uranus next to it has meant the heart has been acting unpredictably the characteristic of Uranus, and physically the emotions have been making the country act, emotions ruled by Cancer/Moon & 4th house.

It's surprising to see as in Australia's chart Pluto in 3rd house (with the same influence as Gemini) the 3rd house has an air element which rules breathing and oxygen and ultimately speech, it also means the illness can be noisy (Gemini/Mercury/3rd house being talkative/mutable element), whilst Pluto means it could be noisy or shallow illustrating the two extremes, making a trine is Venus which softens the abruptness of speech when in the 8th house, ruled by Pluto in another part of astrology 8th house shows where fortunes can be made; the aspect of Pluto can be seen not only in the progressed but the natal chart where over 240 years since independence the progressed Pluto has moved to Aquarius whilst the house position is the same, 3rd house. This can be seen in more than 1 occasion making an influence to the Sun☉/heart is causing the health problem. Pluto♇ is the ruler of death, physically reproductive system and hormones which is ultimately causing a problem with the heart primary cause of heart attacks. Progressed Pluto has just moved on it's an extremely slow moving planet so the focus could still be on this area for some time. Pluto being after the fact would mean the hearts actions would be stopped and only if a person has a chart affected by this.

Cancer

In illnesses Cancer is the 2nd highest cause of death; the Moon combined in the aspects is the contributing factor, whilst it makes for a very caring country it is the constraints that are the problem, and so in illness appropriately placed the Moon is the ruler of the Cancer sign. Cancer is the worrier sign, it also rules the mother/ mothering / memory & home element in our charts, it also harbors resentment, as we know by the animal, when the Crab feels threaten in will retreat into itself. This is what makes it hard for a person to overcome the Cancer illness. In the normal realm of things Cancer is opposite Capricorn squaring Libra and Aries. The constellation making 150-degree angle prior is Aquarius ruler Uranus. This means the person with a Cancer sign would go for is a studious, serious, ambitious even - type person, and if the partner strays there's almost immediate problems.

Where illness is concerned this would only be a problem if a person has their personal chart affected i.e. Moon squaring/opposite or making a 150-degree angle

The Flow Chart is an excellent tool to show how one planet will flow onto another, it shows how progressed planets flow onto both natal and progressed planet and vice versa. When you look at the Flow Chart you will see I've only put in the adverse health aspects where there are considerable aspects being made to the Sun. The aspects seem to repeat the same planets whilst in different constellation.

The Sun and progressed Moon being in Cancer means both women/mothers as separate to single woman, and men would like to be near the water or beach or physically benefit by it.

Natal Mercury in Cancer means their thoughts are sensitive which are disrupted by aggression squaring the Progressed Sun in Aries physically this means that water and fire would be affecting the heart. There is also a square from progressed Sun to natal Moon, this would show a heating up of the stomach juices.

With the 3 planets making a quincunx ⚻ to Pluto, means it would stop the proper breaking down of food, the stomach is what the Moon rules, it would also be a bit hot being the same degree as the Sun, but in a years' time the Moon would have moved 13 degrees so it's specific for this date 4 July 2016. The natal Sun being conjunct progressed Moon and Uranus means the heart would be working unpredictably and Sun and Moon in 9th house means there would be an extension of the stomach and heart region, with the Sun moving onto Pluto it would mean the heart would have a problem functioning properly.

Stroke

The following I believe would show why stroke patients are the 4th highest. Progressed Pluto in the 3rd house would be making a blockage to air, when air is trying to be pumped to the lungs/brain and the ankles. In another area regarding the air element is whilst it doesn't make a bad aspect to progressed Mercury ☿ its influence will flow onto the placing of Uranus ♅ in Gemini, in the 8th house, showing halted speech/excitable communications in reference to real estate and their real estate overseas. Uranus shows unpredictable air flow to the brain and lungs.

Saturn ♄ rules bones/marrow & knees and its the building blocks that cause blockages, this is squaring Jupiter ♃ shows blockage to the liver. The liver assimilates alcohol/drugs and toxins including medication that enter the body, it produces protein that is required for blood clotting and albumin required for circulation, it also turns glucose into glycogen, it makes the bile for food digestion; as such Saturn could cause this area to over work itself and cause silica to go into the bones.

Diabetes

Progressed Jupiter ♃ is also square to natal Venus ♀ ruler of kidneys. This means the breaking down of medication/toxins would be affected by the action of the kidneys. Jupiter is expansive and Venus rules sugar and working in a negative manner is one of the reasons for diabetes, but, more relevant is that natal Moon ☽ and Pluto ♇ is making 150-degree angle to natal Venus.

Alzheimer's

There is also a square between the conjunction of the Sun-Moon-Uranus with Progressed Neptune in Libra, where Uranus is the closest, meaning groups can be confused after an event. Neptune square can give a false reading. I'm not sure not being a doctor how this will work I can only explain it as I see it. Uranus ruling circulation in the Cancer sign, referring to stomach, has a problem with the lymph around the Kidneys, Libra also rules the veins and picks up carbonized blood, Neptune ♆ in Libra ♎. It then flows onto what I explained with Progressed Pluto in Aquarius. This might go unnoticed with the North Node in Leo which in health mainly focus' on the heart, which wouldn't even consider problems with kidneys but inconjuncts also give problems with magnesium/calcium's and salts, Neptune square Moon show complete confusion of memories and early upbringing, this could explain the Alzheimer's disease.
In all cases the degrees are important to show which planet will be blocking the action of the prior especially when they are very close, if the time is wrong it could be working in the reverse.

Below are the planetary symbols for each disease or cause in the case of suicide, for your own reference -

Heart disease	Sun/☉
Cancer (malignant neoplasms aka tumour) -	Moon/☽
Chronic lower respiratory disease	Mercury/☿
Stroke (**cerebrovascular diseases**)	Sun-Mercury/☉☿
(Arteries supplying oxygen to brain blocked)	
Alzheimer's disease	Sun-Neptune/☉♆
(Problems remembering)	
Diabetes (diabetes mellitus)	Venus/♀

Influenza and **pneumonia**	Saturn/♄
Kidney disease (nephritis, nephrotic syndrome, and nephrosis)	Venus/♀
Suicide (intentional self-harm).	Uranus-Sun/♅☉

Britains Health Aspects

I'm taking a punt in the dark by going back to 1689 but it's when the Royals lost or volunteered their power to pass taxes and pass legislation. It was a time when James II fled and there was no Royal family, who had not that many years earlier won, after Cromwell. I've created Britain's Natal Chart for 13 February 1689, when William of Orange and his wife Mary Stuart became King and Queen of England, after Mary's father James II had fled to France. Mary was the eldest daughter of James, Duke of York and King of England and cousin to William. They were married 1677 some 12 years before the countries request for them to become King and Queen of the U.K. Initially the marriage was intended to repair relationships between England and the Netherlands. It was in 1688 the parliamentary opponent to James II invited William of Orange to take the Crown and was assured of the countries support. The extreme importance of this date was that Parliament had passed a Bill of Rights which prevented Catholics for succeeding to the throne and the most extensive as mentioned, law was to limit the monarchs so they could neither pass laws nor levy taxes without parliamentary consent. Ref. http://www.britroyals.com/kings.asp?id=william3 This law has not changed to this day.

I've had considerable problem in finding the approximate figure of the population affected by health problems in the U.K. but, finally by going to the Office of National Statistics (which first said I'd have to pay for it, which I'm totally against when it should be free under the U.N.) and when I finally put in U.K. Bureau of Statistics and again being referred to ONS I then began working my way through referred sites I finally came up with the actual amount of people who had died and by what disease.

Currently England has just over 56 million people as with Australia and the U.S. it has heart disease at the top of their scale yet cancer as a whole surpasses it by 16,000 where I've found it divided into specific areas; after heart disease is Dementia. The following gives the distribution as to the male and female population affected, this is interesting as uneven house/constellation Aries are said to be male and even house/constellations are said to be female, there are also other areas that determine sex but it's not an area I've spent much time on, my understanding is the more planets in a female house the more receptive and non-aggressive they are, I've also gone into it further in -

Leading causes of death for males and females, 2014

England and Wales

	Code`	Deaths Male	Deaths Female .	
1. Ischaemic heart diseases	120-125	36,319	24,190	60,509
2. Dementia and Alzheimer's Diseases	FO1, FO3, G30	17,177	32,321	49,498

3. Cancer - Malignant Neoplasm aka Tumour of Trachea, Bronchus & Lung	C33, C34	16,959	13,909	30,868
4. Chronic lower Respiratory Disease	J40-J47	14,565	14,467	29,032
5. Stroke - Cerebrovascular disease	160-169	14,194	19,963	34,157
6. Influenza & Pneumonia	J09-J18	11,242	14,212	25,454
7. Cancer - Malignant Neoplasm Aka Tumour of Prostate	C61	10,153		
8. Cancer - Malignant Neoplasm Aka Tumuor of Female Breast			10,097	
9. Cancer - Malignant Neoplasm (Tumour) of Colon, Sigmoid, Rectum and Anus	C18-C21	7,718	6,569	14,287
10. Cancer - Malignant Neoplasm (Tumour); Stated or presumed to primary Of Lymphoid, haematopoietic & Related tissue	C81-C96	6,454	5,025	11,479

Cancer

As I've mentioned the main disease in England and Wales is Cancer, and from the wheel below we need to look for three outright areas that cover Cancer and they are the Moon ☽, the Cancer constellation ♋ and the 4th house. I'm a little bit wary of relying on the house solely because it moves every 2 hours and being just over 300 years ago it could be questioned that I've put in the right time; and, I'll find it hard not to refer to it because there is 1/3 more you can find out.

Now I'd like you to guide your eyes to the wheel of William of Orange and Queen Mary, where in the 6th house, you will see in both the inner and outer wheel in particular Saturn ♄ conjunct Moon ☽ in Scorpio ♏. I've spoken in some depth about the Pluto influence and here we see the constellation it rules Scorpio with both Saturn and Moon in it. The things I haven't mentioned is that Saturn rules the oooIlIder generation usually males, and a long or short life would be Scorpio. When we see this together we can say older man and married woman (Moon referring to married women or women with children) in Scorpio refers to the highest spiritual belief with health problems squaring Jupiter in Aquarius (philosophies in groups) and opposite Uranus circulation in Taurus. Saturn and the Moon in Scorpio would mean death by Cancer. The Moon in Scorpio constellation also refers to diseases i.e. colon, bowel, anus, prostrate, uterus, testes, ovaries. Now to add to the problem of knowing which is the most important there is also Pluto ♇ natal and progressed 23 and 25 degrees in Cancer♋. With the house position I would be confident in saying the lung would be affected by the Cancer signs action with the hydrochloric acid. Pluto has nothing in the opposite house but making 150° angle is Mercury at 22 degrees this would mean that thoughts are being blocked by family and that the air/oxygen would be going into the stomach as Mercury is in Aquarius it would be in an unpredictable manner. These aspects give a good reason why Cancer is so varied and high in England and Wales causes of death.

Heart Disease

With heart disease we are looking to the Sun ☉, it's constellation Leo ♌ and the 5th house, when we look to the Royals chart we see the natal Sun ☉ 25° 32' Aquarius ♒ making 150° angle to natal Pluto ♇ at 23° Cancer ♋ and progressed Pluto at 25° 34'. This so clearly shows the health aspect I've spoken of, Sun in Aquarius means groups and circulation this would be blocked by powerful families, the fluids of the body/the electrical nervous system and salts would be heated up with the Sun usually giving warmth but the fluids would be blocked in the stomach, Pluto would add to the stomach calcium and Sulphate which adds to the hydrochloric acid; naturally fluids are required for the blood cells and the muscles to move the fluids around the body and as the body is made up of 50 - 75% water it varies from young to elderly, it is crucial for our survival. In going to other aspects we can see natal Uranus ♅ 19° Taurus ♉ and 300 years later Uranus progressed has only moved to 23° 21' ℞ Taurus; remember progressed planets are calculated 1 day/year so we'd look at 300 days meaning Uranus would have moved 1/7th of its 7-year cycle through a constellation. This being square and the degree's prior natal Sun means conservatives and fixed groups where the health aspect would be an unpredictable thyroid and its hormones would be blocked by the circulatory system Sun in Aquarius, when planets move disease move.

Dementia

Dementia is long term and often gradual decrease in the ability to think and remember that is great enough to affect a person's daily functioning. Other common symptoms include emotional problems, problems with language, and a decrease in motivation. A person's consciousness is usually not affected. A dementia diagnosis requires a change from a person's usual mental functioning - as found by 'Smart Lookup' referring to en.wikipedia.org. Going by the definition we know to look for a person's actions, the Sun ☉ refers to the 'self', the ability to remember and the ruler of the emotions is enveloped in the Moon ☽. Problems with language has to do with Mercury ☿. The ability to be confused is ruled by Neptune♆/Pisces♓ and 12th house.

Now let's look to the wheel Mercury ☿ 01° 03' in Capricorn ♑ this causes a person or in this case country to work at communications and the ability to speak, there had been 150° angle between progressed Pluto♇ and Mercury for some time but as we can see it's gradually moved 5° away. The aspects surround Mercury also shows the importance of timing ruled by North Node NN☊ in the progressed chart it is 00° 05' it emphasis's particular areas and shows a square to Mercury, I often refer to it as having a Jupiter affect and often use the same ingredients when doing formulas. When square it means the country is untimely at speaking. This can be looked upon as an illness, when talking at the wrong time, but it is a progressed aspect so it doesn't have a long lasting affect as Mercury moves faster than the Sun. If we look at the inner wheel showing the natal chart North Node NN ☊ is prior to Mercury in a sextile 60o angle this is a positive aspect.

Where it concerns the emotions and for 2014 the progressed Moon in Scorpio, which is fast moving is trine Pluto in Cancer it would make the emotions very strong and fixed, it squares natal Mercury. This would mean the country is fixed with the thoughts and memories of death and being square would prefer to forget but are being haunted by them. Whilst natal Moon at 6° 32' are almost 150° apart from progressed Mars 02° 57 this refers to the actions being too aggressive in an emotional sense, the natal Moon is in the constellation Sagittarius meaning the emotions are noted overseas as joyful with talking of memories and aggressive as Mars moves on and when it moves away the emotions will less aggressive.

The natal Moon squares Neptune which cases the emotions /memories to become confused as to what has happened overseas/foreignors. The Moon not only refers to the emotions but to mothers or married women or people who have a very caring natures, which could explain the higher rate of death in woman than to men.

Moon in Sagittarius would mean calcium and fluoride in the liver. Heat would be in the thyroid brought about by potassium, Mars ruler of Aries tissue salt is Potassium Phosphate mixed with thyroid hormones mineral rulership nat sulph, this would have a negative affect on the liver. The Moon square Neptune means the calcium and fluoride of the Moon would be combined with the close relationship to Saturn with the mineral salt calcium phosphate, flowing into the feet /Lymph ruled by Neptune, where both natal and progressed Neptune 10° 02/10° 56' ultimately the calcium fluoride would be causing a problem with ferrous phosphate of the lymph. I believe this gives a good explanation to dementia if a person had a negative chart to that of the U.K.

Britain 1689 Natal/Progressed and Transits 2014 Flow Chart

Natal/ Progressed & Transits	Degrees & Minutes	Constellation	House	Planet	Aspects
P	00 05 ℞	Aries	11	NN	
T	00 12	Sagittarius	6	Saturn	Saturn sq Jupiter Saturn sq Neptune
N	00 45	Capricorn	7	Mars	Mars sq Venus
P	01 03	Capricorn	7	Mercury	
P	01 44	Pisces	10	Jupiter	
P	02 57	Taurus	12	Mars	Saturn 150 – P Mars
N	03 37	Aries	11	Venus	Venus sq T Sun
T	03 44	Capricorn	7	Sun	
T	05 14	Pisces	10	Neptune	
N	06 32	Sagittarius	6	Moon	
N	07 04	Scorpio	6	Saturn	
N	10 02	Pisces	11	Neptune)-	
P	10 56	Pisces	11	Neptune)	
T	12 35	Aries	11	Uranus	
T	12 57	Capricorn	8	Pluto	
N	13 13	Aquarius	10	Jupiter	Jupiter ⊼ - Saturn/Moon
T	13 29	Capricorn	8	Mercury	Mercury sq NN
P	15 33	Capricorn	8	Sun	
T	15 40 ℞	Libra	5	Nth Node	
T	16 04	Aquarius	9	Mars	
N	16 24 ℞	Aries	12	Nth Node	210 NN - Saturn; sq Sun
P	16 49	Scorpio	6	Saturn	
P	17 31	Scorpio	6	Moon	
T	18 44	Capricorn	8	Venus	
N	19 02	Taurus	12	Uranus	□ Uranus – Moon/Jupiter/Mercury
P	20 20	Sagittarius	7	Venus	210 Uranus /Venus
T	22 08	Aquarius	10	Moon	Moon sq Jupiter
T	22 10 ℞	Leo	4	Jupiter	
N	22 54	Aquarius	10	Mercury	⊼ Mercury - Pluto
P	23 21 ℞	Taurus	12	Uranus	□ Mercury - Uranus
N	23 21 ℞	Cancer	3	Pluto	□ Pluto - Uranus
N	25 32	Aquarius	10	Sun	⊼ Sun - Pluto
P	25 34 ℞	Cancer	3	Pluto	150 Pluto - Mars

Blue = Natal, Red = Transits, Green = Progressed

Britains Natal and Progressed Chart

CHART 6

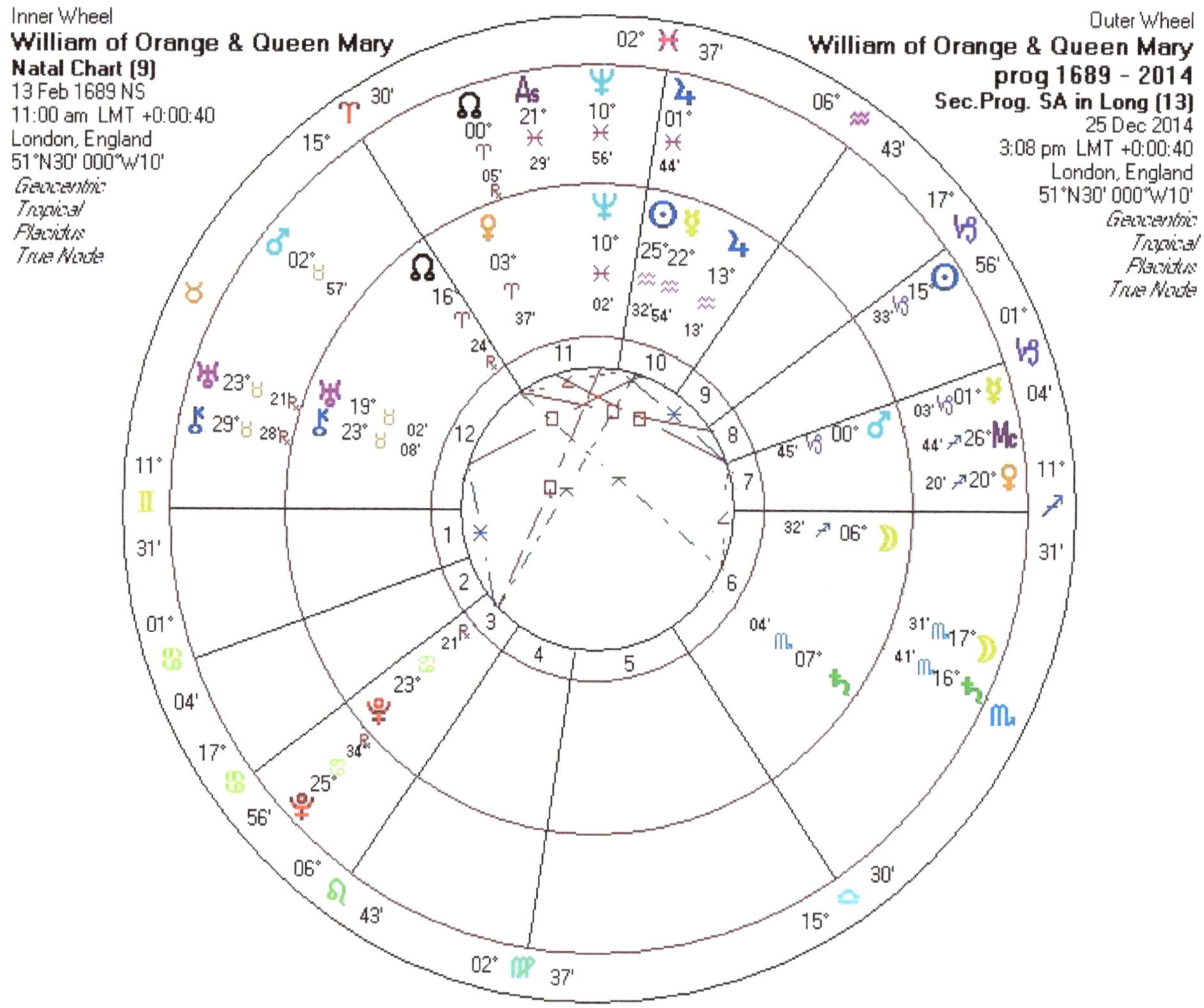

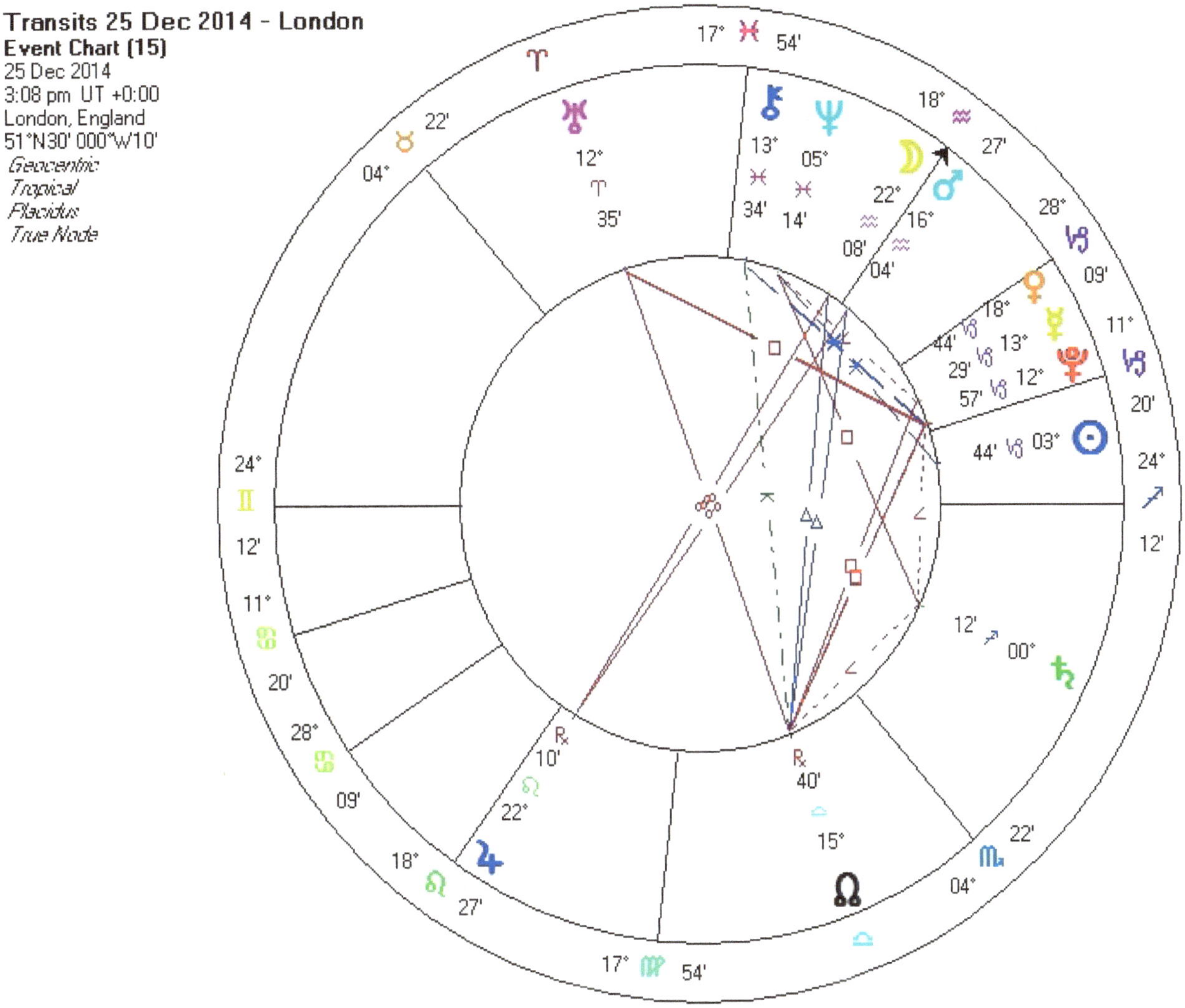

CHART 7

ALTERNATE AREAS OF DISCOVERY

We can also look at the aspects in an alternate manner the way an Astrologer would look at a personal chart but in a countries chart we're looking at it on a broader scale, never the less accurate in America's chart we see Pluto rules the masses this is in the 3rd house, Uranus rules population by way of groups in the 8th house, 3rd house rules neighbourhood and 8th house rules masses as its constellation is Scorpio with the same influence as Pluto and 8th house. 9th house and 3rd house is in opposition meaning there can be excesses, both by foreignors and in the neighbourhood, but married women are very powerful in the neighbourhood Moon conjunct Pluto, with a blow out of population, which is in a sense a health problem; it gives an understanding to their population reaching 320 million at the start of 2015, this is also increased with an easy flow between P. Neptune in 11th house meaning it happens without realizing it,

Pluto in 3rd house of Aquarius and Saturn in 11th house of Libra these are easy flows to increasing the population and particularly those who are married. But, in particular when the Sun and Mercury is in Cancer – means at the time of constitution foreignors found it easy to make America home 9th house – foreignors. If you reflect on the Astrology Chart 3 you'll see by the inner wheel the aspects I've mentioned. You must consider each house covers 30-degrees; which is given a constellation and the planets degrees can be checked by NASA's Scientists Ephemeris

The affects of Pluto is why I'll never consider Pluto as a minor planet, Astronomers can; but, Astronomers do not hold the same faith of Astrologers. Pluto illustrates where you can find death and with the spiritual side of t remake or rebirth, in some respects it supports the promise – death and reincarnation. But death can take many forms e.g. death of finances, death of family life, death of marriage, death of philosophy in fact if you go through the 12 zodiac it can mean any one of the area each cover. It just so happens that in America's chart there are ill aspects to the Sun both progressed and natal chart, which mean people who have conflicting Sun in other constellation making an aspect to the country flow chart would need to be conscious of what to look for e.g. Heating of the stomach, fluid in the ankles, burning of the juices to the heart otherwise a feeling of heart burn. Heat throughout the back

Whilst this book is on Health Formula's and how to create them, readers might be enthused to know that I've have taken it another degree, separate to looking at personalities and looking into the future, I have created several environmental formulas. As I didn't have facilities to prove my theories the last formula I requested reviews through the Environment and Scientific Minister Sen. Kerr in 2007-8-9 my formulas were based on Carbons for disabling Uranium. To this date I haven't checked if they've been used.

Most of my initial formulas on the environment I would find patented after requesting reviews, usually 2 years later, by an unknown source. One I still require information on was on eliminating carbon causing Ozone depletion. I asked John Howard, Australian Prime Minister to request NAZA put it on the MIR Satellite a Crystal. Where I witnessed an Australian Scientist put a crystal on the Mir Satellite. I've been told diamonds were found surrounding it when it came down. When requesting a copy, P.M.'s office said they didn't scan letters in the late 90's and my formula would have been forwarded on to the appropriate dept., which they didn't keep particulars of. Would you believe because I was doing investigations into ITSA, I found a lot of unrelated emails/faxes forwarded to them. This is actually a form of corruption seen by Pluto in 3rd house quincunx Saturn also the ruler of governments.

In addition, the blockage for inventions ruled by Uranus is seen with Uranus ♅ ☍ opposite Pluto ♇ it's not to say this won't change in 100 years unless some drastic change is implemented to the Constitution, with Uranus ♅ progression, but Australia would also have to be careful of written works or from an environmental view point air corrupting what works are in progress, with Pluto in opposition means obsessions and with an obsession it can block a true vision. The outer planets are slow moving, said to be malefic and we have to also take into consideration the Retrograde ℞ aspect, planets that appear to be going backward.

Australia's chart also shows blockage in communication, where deep thought has been ignored by universities where their philosophes are on Plato and how you can relate the written word where someone else has written it. Constraints of Government have added to a gravelly course language. Corruption of ideas is obvious when there are so many planets opposite Pluto a planet that can display the highest good, and rules Popes and great deeds.

Something I've found so relative is with our somewhat abrupt language the allowance of f..k, sh.t, c... in anatomy these areas point to one part of the body ruled by Scorpio/**Pluto**/8th house which you can find prevalent in Australia's chart in communications; and other profanities, our coarse language is always noticed in American movies and is due to the fact that speaking correctly is not of utmost importance in our local schools, manners are also not taught or shown, this would be Venus trait and opposes Pluto 3rd house. These influences are in our Natal chart which means it's a permanent part of our make-up. So it's understandable that at a time when the country wanted course language to be banned swearing was so imbedded that our laws changed from no swearing because of our gaols being over populated to, if it is accepted in the community you're in, it is not illegal. In parts of Australia their trying to change the laws again.
I've diverted mainly for sceptics, but a true sceptic's mind can never be changed unless they can prove it themselves ☺.

Chapter 2 - The Medical, Physiology & Anatomy Connection Between the Close Constellations

The importance between the close constellations prior and after have gone quite unnoticed in astrological texts, yet I first came across the importance of this when making formulas and using the quincunx. I realised it made a figure 8 and in maths the figure 8 stands for infinity. This is not to underrate Square and Oppositions. The quincunx angles have a underlining influence, whilst the square and opposition have obvious influence. I also found that the sign prior has an alleviating influence which is why I've written the chart below.

A few years ago I had cause to put the theory in practice for myself to a doctor in the emergency ward who was about to release me after I was admitted with high blood pressure and it had come down. As soon as I got up I realised I wasn't well at all, I couldn't stand. The pain was causing me to have laboured speech where I suddenly found the doctor actually listening to me. I informed him that through astrology the pain in my knees was caused through a problem with oxygen and the heart. I omitted that is was also through circulation.

My brain doesn't always tick over to put things into a layperson's term and I was as much amazed, as he was when I came out with it. The information allowed him to do the correct tests to find the problem. I was ecstatic. 1. That the doctor understood what I was saying and 2. finding a diagnosis from the symptoms were almost immediate. Later that day when I was home it was in the Australian news that a scientist in England found a sports person had heart problems, which caused failing knees. Gee I'm almost over it. The amount of times I'm researching something, maybe send it off to a Minister to get government review and proof to my research, to find someone else patents it or claimed recognition for it has been never ending, what's more annoying is the Patents Office is allowing whole area and not a specific area to be patented.

But it's a story of untimeliness. Hopefully this will be timely for someone. Distraction is also another problem, so let's get back to it. ☺

To explain my thoughts in telling the doctor that oxygen and heart were involved in causing my knees to fail, I'll go to the quincunx angles in the diagram below. In my case it was the Knees which are Capricorn ♑. The result was that they found I had low oxygen pressure in my arteries. The test is not usually recommended as apparently it's dangerous but I agreed. I needed to know. It was called the Rapid System Text, for myself, I'd say it might be invasive & dangerous if not in a hospital, but you're in the best place if you're going to have heart failure or a stroke. The Dr. punctured me 8x before a registrar helped. It was hard to believe when Dr's said I had to go back to emergency for a check-up they told me it was too invasive. Another reason I look into my own formulas.

In the following graph I've begun with structuring the constellations by their symbols, food types, cell salts and area of the body they cover and then the second column covers the previous sign that will alleviate the problems of the next. In other chapters I've gone into details of others areas that can cause health problems. So remember it's only when there are planets in these signs are they set off by other

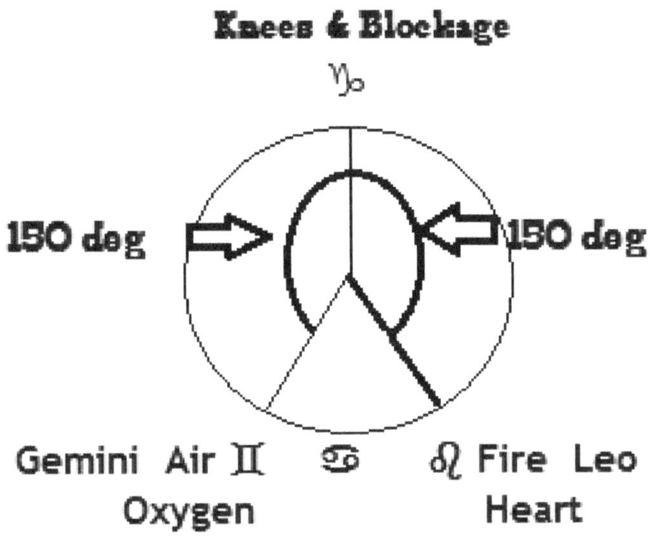

influences. Formulas ren't usually made with one ingredient. And, if you ever hear a person say 'I don't like say Pisces because it's opposite my sign of Virgo' please remember that usually in making a formula I always need the opposite or squaring signs ingredient to complete it.

By recognising the 150 /210 degree importance is what has caused me to put together the influence of the close constellations.
I've gone through the constellations the first columns going down in order beginning with Aries. The second column is showing the influence of the sign before, Pisces - where the theory is the sign before food can help prevent the next constellation prone to have illness.

Physiology & Anatomy Connection Between the Close Constellations	
Aries - Planet Mars Symbol - Ram/Male Sheep Other Names - Lamb>Hogget>Mutton Lamb is high in Zinc Anatomy/Phys - Muscles/Head/ Cover for the Encephalon (Brain)/Red Blood Cells Action - Adrenaline and production of blood Tripl/Quad Element - Fire/Cardinal Medical assoc. - Iron is required for good muscle building and esp. for red blood cells a necessary balance is required between white & red blood cells. Metal - Iron and Steel Lucky Gemstones - Bloodstone, Diamond Jasper, Topaz, Ruby	Pisces - is the sign prior to Aries The Symbol Fish helps to prevent muscle disease. Associated Foods - Most Fish but the fatter and redder the better. Fish have unsaturated fat, different to saturated fats found in butter and those known as saturated fats.
Taurus - Planet Venus Symbol - Buffalo/Bull/Male Cow Other Names - Cow/Veal/Calves Anatomy/Phys - Thyroid/Parathyroid/Neck Action - Thyroid produces hormone that goes to the pituitary gland in the brain, it's function is to regulates fat. Tripl/Quad - Earth/Fixed Medical assoc. - Iodine is required for good Thyroid production and found in food where the iodine will be taken up by the thyroid. Metal - Copper Lucky Gemstone - Emerald, Diamond, Zircon White Coral and Sapphires.	Aries - is the sign prior to Taurus The Symbol Ram helps to prevent Thyroid disease, has no carbohydrates, it is a good source of protein, hemme iron & zinc, also rich in all Vit. B's and foilate. Whilst the meat of cattle has high iron content, the two work together.
Gemini - Planet Mercury Symbol - Twins Anatomy/Phys - Lungs/Arms/Brains/Chest Respiratory & Nervous System.	Taurus - is the sign prior to Gemini. The Symbol the Buffalo will help to prevent lung disease. The Calf -Veal is high in iron potassium and the Thyroid

Action - to put oxygen into the red blood cells, and pick up carbon dioxide to be exhaled. Tripl/Quad - Air/Mutable Medical assoc. - Air is filtered when inhaled and purified in the lungs which creates oxygen. Oxygen is required for good brain function. Metal - Mercury Lucky Gemstone - Emerald, Agate, Pearl and Topaz	hormones are released into the pituitary gland in the brain this helps T3, and T4 to be released into the arms.
Cancer - Planet Moon Symbol - The Crab Other Names - Crayfish Anatomy/Phys - Stomach/Breast (Mammary Glands) Memory/Memories Action - Hydrochloric Acid breaks down the food from the digestive track Tripl/Quad - Water/Cardinal Metal - Silver Lucky Gemstones - Opal Pearl, Ruby and Moonstone	Gemini - is the sign prior to Cancer. The Symbol is the Twins. It's quite likely the Gas Mercury - Mercuric Chloride is the substance found in the lining of the lungs that tells the brain what is good and bad. The Brain then informs the stomach when to purge and to continue processing the food.
Leo - Planet Sun Symbol - The Lion Affiliation with - Tiger/Leopard Anatomy/Phys - Heart/Solar Plexus Trip/Quad - Fire/Fixed Medical assoc. - the heart is a vessel which has a close reliance on the lungs. The heart pumps and deoxygenate the blood for the lungs to oxygenate Metal – Gold. Lucky Gemstone - Ruby, Sardonyx	*Cancer is the sign prior to Leo. The Symbol the Crab. Crab is a natural source of omega-3 fatty acids, which decreases chances of heart attack.
Virgo - Planet Mercury Symbol - The Virgin Other terms - Virginal/Unspoiled Anatomy/Phys - Small Intestines	Leo is the sign prior to Virgo. The Symbol is the Lion. Leo rules the heart and it's believed the juices and meat from the heart of an

Trip/Quad - Earth/Mutable Medical assoc. - Absorb nutrients and minerals from food. 90% Digestion occurs in the small intestine compared to 10% in the stomach. Contains pancreatic lipase, in duodenum then sucrase, lactase and maltase that break down smaller components. Metal - Silver Lucky Gemstone - Sapphire, Sardonyx	animal would help intestinal disease.
Libra - Planet Venus Symbol - Scales Anatomy/Phys - Kidneys aka Renal Track and Urinary System. Trip/Quad - Air/Cardinal Associated Area - Balancing the emotions balancing electrolytes in the blood along with balancing ph homeostasis. It also removes excess organic molecules from the blood. Metal – Copper Lucky Gemstones - Sapphire and Turquoise	Virgo is the sign prior to Libra The Symbol is the Virgin The eating of intestines which is called chitterling or as Balkans, Greece and Turkey call it Kokoretsi/Spain - Gallinejas could be helpful for kidney problems. Usually these are animal intestines, cooked and prepared at specific times.
Scorpio - Planet Pluto Symbol - Water Scorpion Associated - Scorpion poison Anatomy/Phys - Penis, Srcotum, Vagina, Anus, bowel and womb in general reproductive system. Birth & Sex. Trip/Quad - Water/Fixed. Metal - Copper Lucky Gemstone - Bloodstone and Topaz	Libra is the sign prior to Scorpio. The symbol is the scales internal organs are the kidneys. As this is a metal it could be referring to Copper. An increase in these could help Scorpion problems.
Sagittarius - Planet Jupiter Symbol - The Archer, on a Horse. Affiliation - The Horse Anatomy/Phys - Liver/Skin/Hips and Upper Thighs. Also Integumentary System Trip/Quad - Fire/Mutable	Scorpio is the sign prior to Sagittarius The Symbol is Water Scorpion. To prevent Liver disease taking of Water Scorpion could help. Not to be taken without medical advice. This is untried whilst dried scorpion has been tested.

Metal - Tin Lucky Gemstone - Topaz	
Capricorn - Planet Saturn Symbol - The Goat Associated - Buck/Nanny/Billy/Doe/Kid or Mutton Anatomy/Phys - Knees/Bones/Marrow Trip/Quad - Earth/Cardinal Metal - Lead, Silver Lucky Gemstone - Amber, Onyx and Dark Sapphire	Sagittarius is the sign prior to Capricorn The Symbol is the Half Man Half Horse To prevent Osteoarthritis the Horses Liver should be trialled to prevent Arthritis, Rheumatoid arthritis, and Osteoarthritis
Aquarius - Planet Uranus Symbol - A Male Carrying Water. The Water Carrier Associated - Electrodes/Electrical Nervous System Anatomy/Phys - The Circulatory System and the Ankles Trip/Quad - Air/Fixed Metal - Aluminium Lucky Gemstone - Turquoise	Capricorn is the sign prior to Aquarius The Symbol is the Goat Capricorn rules the knees. Too much air in the Electrical Nervous System could benefit from the Marrow particularly in the Goat's knees esp. noticed when boiled. This could be rubbed onto the affected parts.
Pisces - Planet Neptune Symbol - Fish Anatomy/Phys - Feet/Lymph System and White blood cells. Trip/Quad - Water/Mutable Metal - Platinum Lucky Gemstone - Moonstone, Cat's eye	Aquarius is the sign prior to Pisces The Symbol is the Water Carrier Electrodes would benefit the Lymph Nodes and White Blood Cells if there

Chapter 3 - Acids & Alkalines

With Acids and Alkalines the uneven numbered constellations in the zodiac are alkalines and the even numbered constellations are the acids as shown below:-

Acids & Alkalines

Alkalines:	Aries, Gemini, Leo, Libra, Sagittarius, Aquarius
Fire/Air	Mars, Mercury, Sun, Venus, Jupiter, Aquarius
Acids:	Taurus, Cancer, Virgo, Scorpio, Capricorn, Pisces
Earth/Water	Venus, Moon, Mercury, Pluto, Saturn, Neptune

All the Alkalines are either Fire or Air e.g. Aries – Fire, Gemini – Air, Leo – Fire etc. Acids are put into the triplicities of earth and water respectively. A simple test to find out if something is alkaline or acidic is by using litmus paper it will change from red to blue for alkaline conditions and blue to red for acidic conditions. In numeric form 7 up shows alkaline, and 7 under is acidic.

There are many alkalines and acids of the body which are used by specific organs of the body (which I would say God created, for lack of another description, but there again I believe God created everything) and each do specific things, our body appears to be like a jigsaw puzzle, remarkably everything has a place and required action, what this book is intended to do is to show a link and give an astrological view of anatomy and physiology, which could be added to veterinary studies. I've noticed in Botanists studies colour can contribute to a poison, but in astrology this could be attributed to the area in which the planet is found.

It's been in the botanist's findings have I found plants dispersing gas, which alternately can be formed in the body due to the acids of the body and through oppositions. For me it's exciting to see in the alkaline, it was the first thing that I came across which had the first relationship with something the medical field could understand. In my formulas I often look for an alkaline I can scour to turn into an acid, because in an example I've given further on, an alkaline such as Mars could be in Taurus found under acids.

Our main gas is oxygen which has no ph. So one might presume it has a 7ph I say this because the Sun/Leo is the precursor to it, and Gemini an air sign shows the lungs which it rules causes the oxygenation of the blood all necessary for every part of the body and is caused by photosynthesis made through the earth's atmosphere and the Sun and found mainly by the trees.

As we are taught each sign has a tissue salt we can imagine each of the oxides that would be associated with the Anatomy of the human, in some texts of astrology the Sun rules oxygen because of the photosynthesis it creates, the planet Uranus rules Nitrogen because it is unstable, Hydrogen is ruled by

Saturn. In the Zodiac for the Liver, with Jupiter being the ruler we have Silica but if we add one more hydrogen to it, it becomes an acid. The acid grouping for hydrogen is Saturn or Capricorn. For Kidneys they require Natrum Phosphate with Libra and Venus being the ruler, but when Saturn or Capricorn is found affecting Venus ruler of kidneys we could find Cirrhosis a thickening of the kidneys caused by a hydrogen, in astrology hydrogen could create bone growth, nucleus is also ruled by Saturn and the elderly where Nucleic Acid is supposed to be the cause of aging; there are the natural foods to replace this i.e. wheat germ, bran, spinach, asparagus, mushrooms, fish, (esp. sardines, salmon and anchovies), chicken liver, oatmeal and onions. A diet of seafood 7 times a week along with milk and juices.

In the acid family we see constellations and planets that have Earth and Water elements. The stomach ruled by the Moon requires Hydrochloric Acid it requires a degree of acid to breakdown the minerals and foods digested and when there is an upset stomach charcoal is taken. Charcoal is a Calcium but an Earth Calcium, it's a product of potash/burnt wood, whilst Calcium is also an alkaline in Tissue Salts of Cancer, Capricorn and Scorpio, one requires a fluoride, the other a planet food phosphate and the last a Sulphate. Calcium can also be a lime; in an association of words this makes me wonder if the Lime fruit could be used.

From understanding acids and alkalizes we will discover a whole new area giving an understanding to scouring food substances. And, an area the goes more into the terms the medical field understand.

Ionizing is where the chemical reaction takes place for it changes one substance to another. In my studies in Astrology it's caused by planets + specific constellations which have to be counteracted and the best way I've found is by foods making an alkaline food into its acidic form when required.

In the above chart we find a planet and 9th house (ruled by Sagittarius) which are alkaline, and a constellation which is alkaline. Carrots are generally alkaline, when you need an excess of acid the procedure informs you by the elements that there is the requirement for heat, salt water and drying, and therefore the produce needs to be boiled and fire dried. Rice has already been dried and could be ground and added later. Alternately rice flour could be used to thicken it.

Alkaline & Acid Food Types

Alkaline Foods (- ph)

• Vegetables – especially *raw* green leafy vegetables e.g. boiled broccoli, black beans – cooked, green beans - boiled, lima beans – cooked, cucumber peeled, raw cabbage, carrots – boiled, mung bean, , kale boiled, lettuce iceberg, peppers – green sweet and raw, potato – baked with skin, tomato red

• Fresh Herbs and Spices –basil, cilantro, cayenne, ginger, parsley.

• Fruits – apples, avocado, bananas, blueberries fresh, young coconuts, dried dates, grapes fresh kiwi fruit fresh, oranges fresh, strawberries fresh, watermelon,

• Wheat grass.

• Sprouts i.e. alfalfa etc.

• Sour cream

Acid Foods (+ ph)

• Junk and Processed foods.

• Sugar.

• All animal food: - meat – hamburger, turkey, eggs, chicken; fish - catfish baked, lobster, oysters, salmon baked, tuna which is chunky light an canned.

• Grains: (White) wheat, rice pasta flour, bread, garbanzo beans – cooked,

• Some Fruits: - pecans raw, English walnuts raw

• Dairy products: - milk, butter, cheeses i.e. Cheddar, Cottage cheese, Cream cheese – regular, Swiss cheese, plain yoghurt low fat

• Peanuts cashews.

Ref: http://www.thebestofrawfood.com/raw-food-diet-conversion-chart.html
http://www.essense-of-life.com/moreinfo/healthtopics/A-701/foodcharts.htm

Chapter 4 -Triplicities and Quadruplicities

The Triplicities and Quadruplicities give a quick reference to how people will act particularly when there are planets in the constellations. Triplicities give an easy flow, whilst Quadruplicities give a harsh flow.

Triplicities Δ - Trines

The four triplicities and groupings are separated by 120°, the angle primarily refers to the constellation but it could extend to referring to houses and planets as planets rule particular constellations, and houses take on the characteristic of the constellations. The four Triplicities are Fire, Earth, Air and Water.

Often an astrologer will refer to a planet not only a constellation as fixed, active or talkative. In western astrology the Triplicities are Fire, Earth, Air, Water. These groupings cause easy flow and each have a Quadruplicit side for when we go around the zodiac we find one is active, the next is fixed and the last is mutable which are the Quadruplicities. The way the constellations flow with Aries ruler of the first house shows if a planet was in that sign it would be active and fiery if there was an affliction by another planet otherwise there could be a warmth and temperate feeling, then when we move to the next easy flow we find Leo, a planet in this house would mean it would be fixed but again if affected the heat of both signs could mean an overheating of the heart, Leo ruler of the heart, this could happen because they mightn't be inclined to act because of the fixed nature, which in fact could be unhealthy, and would probably be more so if it was the Sun in Leo, the next trine is Sagittarius which is mutable/talkative. Planets can make an easy flow to these trines and it is important to understand the temperament of the triplicities. People of the same Triplicities find it easy to either speak, act, be fixed or feel emotional attuned to another person's feelings, water element.

The Triplicities of Constellations, Planets & Houses

Fire Signs –

Aries, Leo, Sagittarius; Mars, Sun, Jupiter, 1st, 5th, 9th house.

Earth Signs –

Taurus, Virgo, Capricorn; *Venus, *Mercury, Saturn; 2nd, 6th, 10th house.

Air Signs –

Gemini, Libra, Aquarius; *Mercury, *Venus, Uranus; 3rd, 7th, 11th house.

Water Signs –

Cancer, Scorpio, Pisces; Moon, Pluto, Neptune; 4th, 8th, 12th house.

Quadruplicities □ Squares

Quadruplicities are groupings of 3 where there are 4 Constellation/Planets and house with same element, separated by 90°, the quadruplicities are Cardinal/Active, Fixed and Mutable, and this affect goes onto the ruling planet and corresponding house. E.g. Taurus, Leo, Scorpio and Aquarius are all fixed signed when they make their minds up not to do something no one will change their minds, but there are exceptions a father might change a Leo's decision, a group of friends may change an Aquarian person's decision, a fraternity may change a Taurian's mind. The other two are Cardinal and Mutable sign. A group of people with all mutable signs may find themselves talking a lot but not taking too much action; this would be alright in a book club; but, not necessarily good in building a house unless you're the architect, that would be the job of the cardinal signs.

The Quadruplicities of Constellations, Planets & Houses

Cardinal/Active Signs –

Aries, Cancer, *Libra, Capricorn;
Mars, Moon, *Venus, Saturn;
1st, 4th, 7th, 10th house.

Fixed Signs –

*Taurus, Leo, Scorpio, Aquarius;
*Venus, Sun, Pluto, Uranus;
2nd, 5th, 8th, 11th house.

Mutable/Talkative Signs –

*Gemini, *Virgo, Sagittarius, Pisces; *
**Mercury, Jupiter, Neptune;
3rd, 6th, 9th, 12th house.

Astrology is best understood in a simplistic view that by associating words / colour / shape / taste / feel / smell etc. to a particular element /mineral / plant / or animal etc. you'll be surprised how you'll find curers. In other words, each planet rules a colour or colours you can then relate this to the colour of the illness e.g. a person burning with a fever and red hot we'd then be looking for Mars in a chart.

It also has been seen that shape can contribute to cures. This could be explained simply by looking to a walnut and the shape that it gives that resembles the brain and it is believed to help the brain function. . This suggestion may not have been confirmed by the medical profession, as yet, but, in astrology it would be one of the foods used in lung and brain cures, which I'm sure would work if used in the right way and for the right purpose. Something I can confirm is the chestnut 'chest' the impressive word. I was given a chestnut once from pharmacist when I was deep with congestion, he told me to boil it and inhale, the result was instant, and I think drink the it as tea. The same could be for the walnut boiling and inhaling it, as Gemini is an air sign and rules the lungs, with the prefix being 'wal' and suffix 'nut' if we relate to the shape being round like the head we could summarize that the 'walnut' could help the lining of the head. Please refer photo.

The spear shape of wheat/ barley relates the product of the planet Mars, constellation of Aries it's the warrior with the spear, also Mars relates to sudden cuts and accidents, whilst the product isn't hot, firing it up (as it's a fire sign) could help with pimples in fact I have found the fermenting of things leads to alcohols which can help i.e. whisky, whisky is made from barley, which I've found invaluable with measles; skin aliments are primarily a Jupiter problem and Gin is made from Juniper plant, Sagittarius ruled by Jupiter would be affected by the Moon or Cancer sign, Cancer making that quincunx degree orb this would require a product to be fermented; the Moon or Cancer constellation rules the stomach and the action of the stomach is to use hydrochloric acid in the stomach to ferment foods. Cancer sign also has dominance over the mammary glands that produce milk. A common cure for upset stomach is Yakult a fermented milk. If you can see a relationship I used to think to myself and the product is used regularly and it hasn't killed anyone then it's worth a try; with using whiskey on the head for measles not only did it relieve the itching, it brought down the fever.

In the association with words and plants or other products, I have found the Latin/Old Roman and Greek often have meaning behind part of the word with its many variances or other cultures, to the part of the anatomy and tell you which plant will help. Relationship can be found in Botanical names, not necessarily in the full name but by separating it into syllables. I believe, what is generally known through the great tombs of Egypt is that the Pharaohs had a belief structure in astrology and the heavens, which gave us our first hieroglyphics

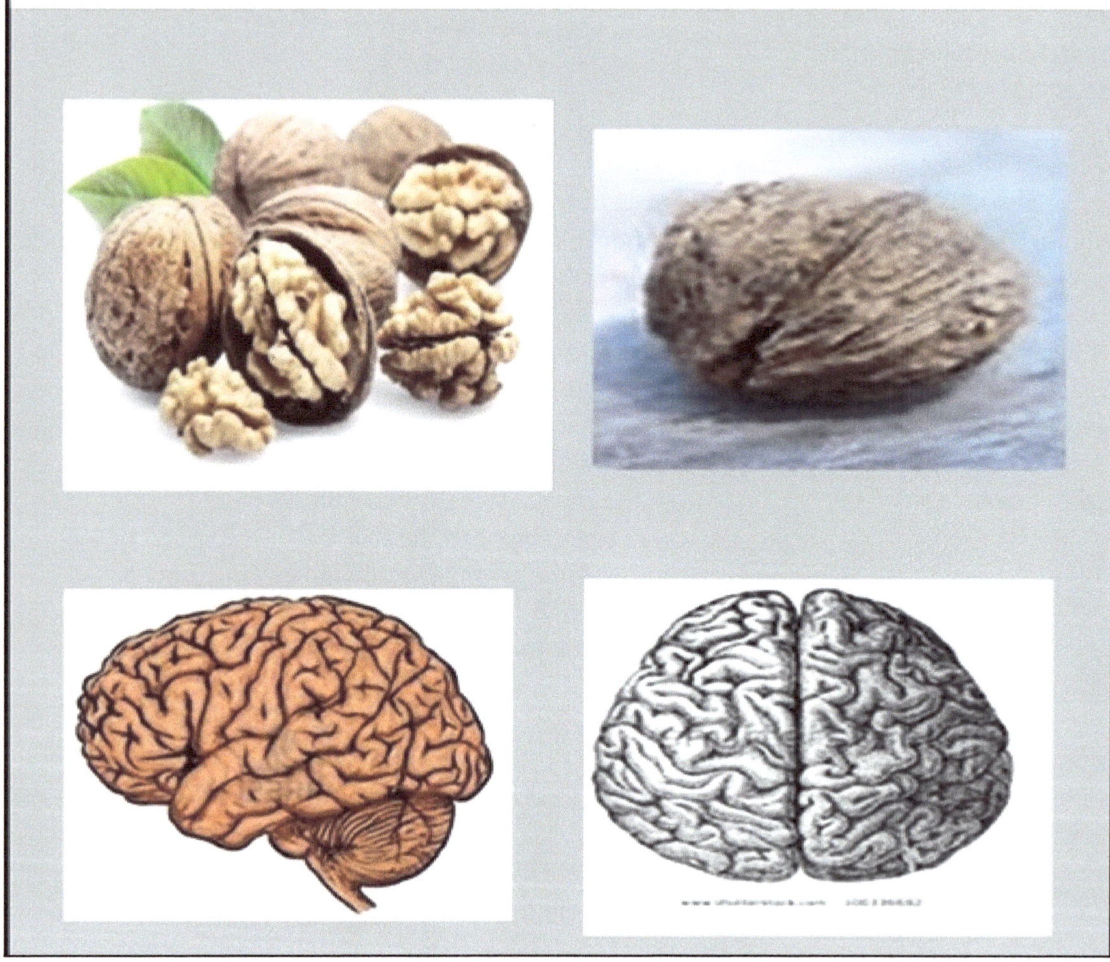
WHAT YOU SEE COULD BE A CURE?

Chapter 5 - Climate

Climate can affect the health, and we don't often think of it being an alkaline or an acid, and yet perspiration is either an alkaline or an acid and sometimes neutral. Climate can change in one day not necessarily by weather but by the environment that we move to i.e. an air-conditioned building.
In saying that a Climate can be understood by planets, it would probably be affected the most by the moon, and planets that are in your chart combined with others. The moon is very sensitive and rules the emotions and as mentioned the stomach, which can be related to the physical body as below we can also see changes in the atmosphere of countries, personally it is not my field but I will try and explain it:-

Sun is warm to hot and fixed. If there is an affliction it could indicate a person with a high temperature.

Mercury is cold, talkative and an air sign in Gemini but, an earth sign when seen in Virgo, which rules the intestinal area. Environmentally it could indicate a wind that is noisy blowing the leaves through the trees, in health it would indicate extreme barking cough and problems with air.

Venus is dry and fixed or moveable (fixed means will not move until some favorable or unfavorable aspect occurs, Venus is so far the ruler of both Taurus and Venus which is why I've stated 'fixed' and 'moveable') it would indicate an atmosphere that won't move.

Mars is hot and moveable - If a country is in a heat wave we could expect a number of fire signs in a fire constellation. The fire signs are under Quick Reference to Elements.

Moon is warm and moveable. - This would indicate a warm humid atmosphere. It is the fastest planet/satellite of earths that affects it most personally. Within 2 ½ days it would make several aspects to Uranus which is the most unpredictable planet.

Saturn is humid and moveable, and would indicate a cold humid atmosphere, in health it would mean someone who feels clammy.

Jupiter is windy, hot and talkative, with wind there would be the underlying factor of Uranus. Jupiter rules beaches, the wind would bring a relief to the hot weather. And, to explain talkative, the thrashing of the ocean on the beach is heard from the sea waters.

Above are the closest planets to the Sun. But along with temperature some planets have an atmosphere full of gases and the combination of these planets gases can help to alleviate an illness. Instead of dissecting the elements such as seen in the periodic table the planets outline the gases they contain this was given initially by its colour or mixture of colours with primitive or not so primitive telescopes that Astrologers/Astronomers put down to the earth's substances they knew gave a particular colour.

How the illness is formed is through a combination of squares, oppositions and other aspects of planets that can be seen by the degree which are near each other, which one could take quick reference using NASA's Scientist Ephemeris showing planets and their degrees. Planets don't usually move much. Calculations are usually required when the different planets degrees are virtually the same. These planets then oppose or square planets that are conjunct (conjunct – at the same degree). These are obvious illnesses but it is the sesquares and quincunx's that complicate any illness, these are 30 degrees and 150 degree aspects respectively. It is important to remember when making a formula the degree and minute shows us which is prepared first and other planets foods are added if close, within 5 degrees.

The outer planets do play an important part in weather.

Uranus is unpredictable it is air sign so we'd expect winds. It is a fixed sign and how this works could mean that the change that takes place could be at the same time. Like the news stating the temperature will change at a certain time from an easterly. This then could be activated by the Moon being a cardinal/active sign.

Neptune is a water sign and I'm not sure if it has a temperature as such, if you think of a ship in the ocean, where sailors can be on their ship for days on end without a breeze. It takes me back to believing it has to be activated by cardinal sign.

The last is Pluto again this is a water sign. Astronomers have downgraded it to a minor planet. And, there is no proof of Chiron having any real influence except astrologer's stating it is a healer, there is some talk as to whether the Sun goes through its constellation which is believed to be Centaurus, and there talk of a clashing of constellations with Scorpio, in a millions years' time... Pluto is another fixed signed and it stands for disaster at its worst, but, what affect humans is not something that is necessarily detrimental to the earth unless a square or opposition to the Sun. Therefore, if Uranus is making an aspect to Pluto in the 10th house we could see unpredictable volcanos causing the earth to renew itself, or the breaking up of the North and South poles because of Pluto being in Capricorn.

What is said about the outer planets is that they have a characteristic, which happens and affects a generation as they move gigantic things when involved, and, their length of time in a constellation can be considerable e.g. Pluto take 13-32 years to go through one constellation.

Chapter 6 - The Influences of Constellations as Compared to the Planet and House

Below is how I've found to analyze a chart, it's the basics, whilst Arabic astrologers begin their charts with Pisces on the first house, I put Aries on the 1st house, as Western astrologers do and have found no problems in finding formulas, but, Western Astrologers mainly analyze personal planets and how they affect constellations and houses.

The below is how I view things both personal and emotional when trying to find a formula. Illness is not necessarily one thing and if you find there is turmoil in one of the groupings you can look to the inner organ to be upset.

PLANETS MOVEMENTS

The planets movements show how turmoil's can move on, as the inner planets move a lot faster than the exterior.

The planets length of time it takes to circle the Sun are shown below. I've included the Sun but it is really as it's seen from Earth. And, the Moon is actually circling the Earth.

Sun – 25.38 days
Moon – 27.321 days through all the constellations from earth

Below shows the time all other planets take to go around the Sun and through the 12 Constellations an ephemeris is required to see in which constellation they are in and the varying length of time they're in it.

Mercury – 88 days, Mercury is only ever 28 degrees away from the Sun and can spend only 4 days in one sign.
Venus – 245 days through all constellations
Earth – 365.24 days through all constellations, it's the Earths view of the Sun that we see Sun move through the Constellation, it'*s the degree that we judge our Zodiac sign or Sun sign, which most people are interested in.
Mars – 687 days
Jupiter – 11.86 years, spends approximately 1 year in each sign
Saturn – 29 ½ years = 10,759 earth days spends approximately 2 ½ year in each sign.
Uranus – 84 earth years
Neptune – 164.79 years
Pluto - 284 days can spend 13-32 years in a sign.
The Chart below shows Constellation/House/Planet and the Triplicities and Quadruplicities that change them.

Constellation	House - Take from TOB	Planet	Quadruplicity	Triplicity
Aries Taurus	1st house 2nd house	Mars Venus (this is sweet earth)	Cardinal-Active Fixed	Fire Earth
Gemini	3rd house	Mercury (as in air)	Mutable Talkative	Air
Cancer Leo	4th house 5th house	Moon Sun	Cardinal - Active Fixed	Water Fire
Virgo Libra	6th house 7th house	Mercury (as in earth) Venus (this is sweet air)	Mutable - Talkative Cardinal - Active	Earth Air
Scorpio	8th house	Pluto	Fixed	Water
Sagittarius Capricorn	9th house 10th house	Jupiter Saturn	Mutable - Talkative Cardinal - Active	Fire Earth
Aquarius	11th house	Uranus	Fixed	Air
Pisces	12th house	Neptune	Mutable - Talkative	Water

By following this you'll be able to relate to the planet and house, and constellation ALSO known as the sign. The Constellation is the area in the anatomy; the Planet is the influence or what's being brought into the area and the house shows the area of the body that is being affected.

The Quadruplicities and Triplicities are elements which are extremely important in analyzing they illustrate fixed or for instance Aquarius is fixed and so is a trunk of a tree, the croton tree is specific to Aquarius the ***Croton tiglium tree*** is medical use is for healing lesions; severe constipation as a purgative, it is used in wind breaks ; In the Amazon the red latex from the species *Croton lechleri*, known as Sangre de Drago (Dragon's blood), is used as a "liquid bandage". Another interesting point is that it's been found to be a source of the organic compound *phorbol and its tumor promoting esters,* analyzing 150 degree angle Aquarius is an underlying problem for Cancerians https://en.wikipedia.org/wiki/Croton_(plant) I should stress a person who's born the zodiac sign of Cancer is not necessarily going to die of Cancer; health problems are determined by ill placed planetary aspects. It generally means that freedom /unpredictable actions play on the home element, when planets are in it. In the 1940's – 50' Uranus was in Cancer; Uranus rules Aquarius with the same influence shows homes 20-30 years on would be experiencing unpredictable groupings in the home where some people could flow with it whilst others wouldn't.

By understanding the elements actions e.g. if Mars was in the 2nd house ruled by Taurus, the person is likely to bring a warm aspect to the area, would be active in brotherhoods, the place is likely to be where there's male or female bonding, you'd automatically be brought to Mars – warmth or hot and active, Taurus is bonding or same sexes. In western astrology the philosophy begins with Aries once you understand their character you will then see that the three areas use different parts of the characteristic to operate.
Another example is to put Cancer on the Ascendant, which is the first house a person with this aspect would show themselves as being caring and creative in any area there in, they might work hard at keeping the place clean, but using the term 'work' would bring in a Saturn influence, but, it has a natural active nature, I should add in a progress chart the houses will change, the planets move...the following is an explanation to this theory...

Aries– 1st House - Mars

The **constellation Aries** is direct as if they know what they're doing, they're an active sign (active meaning cardinal, and, fiery meaning instantaneous) it has a natural warmth coming due to its fiery element; it gives the want for movement, it's pointed in shape. Its keyword is 'I' The Arian nature gives Taurians the reason for doing what they do. Aries rules the part of the anatomy which is muscles throughout the body and those of the skull and muscle surrounding it. External – the Head.
The **1st house** rules the environment it would be hot, a person's likely to be living in an arid area, may have a pointed roof, possibly red brick, or geographically maybe where wheat is grown or even thistles. In medical terms their head could be hot or whole body. It's the Planets and Constellations that will change it.
The **Planet Mars** shows the influence in the area it is found, Mars is hot, active, i.e. if the planet is afflicted - temperature would go up and down and up again, if afflicted – inflamed and red

Taurus – 2ND House – Venus
(Venus the Earth element as opposed to Libra)
The **constellation Taurus** is fixed and persevering, and earth sign, being fixed means immovable, so when they decide to stop doing something they will not change their minds unless they decide to. Yet their influenced by their male or female counterpart. It rules money, food, possessions, conservative ideas, their conservative ideas are what generally is accepted by those they grow up with. It is also a sign of brotherhood of the same sexes e.g. Freemasons or Country Woman's Association (CWA). They have the ability to continue the Arian dream. Keyword 'I have'. What they do in today's society leads to Gemini. part of the anatomy is the Thyroid/Thymus and the growth hormone, the pituitary gland, this also regulates blood pressure and stress. External- the Neck/Throat
The **2nd House** rules the environment, the area would be the throat and pituitary gland
The **Planet Venus** is to influence the constellation it's in by the influences of the constellation it rules, medically it could cause the constellation to be sweet, with sugar. As it rules the throat if Venus was in Taurus it would mean a very nice voice.

As an e.g. Mars in Capricorn 2nd house, there would be a lot of action, heat and constraint happening in the thyroid and pituitary gland. Both Mars and Capricorn are active their functions are opposite each other, their colours are red, black and pink.

Gemini – 3rd House - Mercury

(Mercury the Air Element as opposed to Virgo)

The **Constellation Gemini's** part of the anatomy is the lungs, brain which rules the breathing process which leads us to the action to push through air and here we can also see the muscles action of the earlier sign in the lung which would be the testosterone, which would not even be thought of. Externally - the arms.

The Planet Mercury is talkative and versatile in their thoughts with an air element. The planet Mercury causes a person to talk on varying aspects of constellation it's in and the aspects affecting it. Talkative meaning mutable and versatile from being an air sign, it will be talkative in whatever area you find Gemini/Mercury, planet and 3rd house. It rules siblings, the neighbourhood, early schooling, transport and communicates/communication, correspondence. Keywords 'I have an idea'. It's anatomy is the Brain and Shoulder and Arms E.g. Mercury in Capricorn prone to talk about work.. It brings air/oxygen & earth into the equation,

The **3rd House** the environment as known by Gemini, of course its physical area would be the lung or brain. If afflicted, you'd notice problems with breathing, speech and with arranging thoughts, but the area these things would be noticed, in the neighbourhood, with brothers and sisters, in educational areas, possibly using different types of transport. It could even be stressful writing maybe to governments or court hearings.

An example would be, if we saw Mercury in Scorpio in 3rd house, it means that air would be affecting the sexual hormone as it's negatively placed, which could be causing problems with communication - 3rd house pointing us to the lungs and brain; two air elements 3rd house and Mercury means excessive air maybe appearing as hyperventilating. Mercury in Scorpio could also cause a person to talk during sex. Or in a medical sense causing too much wind with too much air; it probably wouldn't be too good for a spy.

Cancer - 4th house - Moon

The Cancer Constellation is sensitive in their actions, and is an active sign being cardinal, and watery. It rules family, babies and children and the home and needless to say 'tears'. It also rules the emotions/ memories and habits. Keyword 'I care' It's anatomy is the Breast/Mammary Glands and Stomach its role is to produce hydrochloric acid that breaks down foods.

The **Moon** brings with it feeling. Its physical aspect is water, it can worry about things and its sensitivity causes it to withdraw if challenged, this would be seen in the area of the constellation. As with it counterpart the Moon would bring into the area hydrochloric acid and only would affect those areas which are square and opposite and making 150-degree angle.

The **4th House** position as with other houses are only affected if there is a planet in them, or those that contradicts the house or constellation it's in. Where there is an upset you'd find it would be noticed in the home or with family, and around babies. A mild case would be heartburn.

Leo – 5th House - Sun

Leo Constellation is fixed and fiery, proud and sensitive (a characteristic formed from Cancer), it rules teenagers, entertainment, the soul/the self, it's royal and can indicate Royalty. Keyword 'I am', without being sensitive it wouldn't notice the admiration of others to feel proud and can lead to being vain, as with myself not wishing to go out without my hair done. People may think this is just a Leo sign, I'd like to explain the Sun which rules Leo can be in any of the 12 signs and where ever, that is, is where a person feels Proud, even if they don't know it. it's known as the zodiac sign, but more of that under Sun. Its anatomy is Chest and the Heart Organ, the atrium, the ventricles and the pulmonary areas, And there is a close relationship with the opposite sign which rules circulatory system. Itself gives out warmth which often goes unnoticed, it's also one of romance, where ever the constellation is in your chart, there's also an interest in helping youth groups, this is mainly seen if there are planets in and making an aspect to it.

The Sun takes approx. 30 days to go through constellations which is the dominant factor in the zodiac. It's royal, it shows pride, it's influence is seen by individual zodiacs more than any other planets. The heart would be affected if afflicted by 150 degree angles by other planets and constellations, but naturally with squares and oppositions.

The 5th House is the area of entertainment and will depict where youth groups meet. It will be noticed if planets are in it and afflicted by other planets. It's also the house of romance. E.g. if the Planet Pluto is in Taurus in the 5th house; Pluto would make the person appear quiet yet powerful, when in the 5th house it would be noticed in groups the opposite sign Aquarius rules; they would be the one to stand out and do things by themselves, Pluto, Taurus in 5th house could mean too much acidity in the heart and hormonal test might show it's being absorbed by the heart where they could be overeating causing cholesterol problems. If there are no other planets and constellations affecting it there might be no problems other, then weight.

Virgo – 6th House - Mercury

(Mercury the earth element, as opposed to Mercury created for inhaling .02mg)

The Virgo Constellation is talkative and prudent in the earth sign, it is also practical and can be critical of themselves and others, looking for perfection, it indicates a way and area in which one works. Keyword 'I can' It is virginal and bring about the attitude of no sex in the work are, it's an area and sign where they want to be of service but it also rules computers and working with small things. Its anatomy is the Small and Large Intestines. As the intestinal area is quiet twisting it is interesting to note that snake venom, snake penis, are offered in Chinese Medicine as good for the intestinal region and for intestinal upset.

6th House is where a person will find the type of work they'd be suited to; it is where they want to be of service. It points to the intestinal region if afflicted.

Mercury the ruler of both Virgo or Gemini will only take on the Virgo thoughts if in Virgo constellation. It is strange that the great Astrologer/Astronomers chose this planet for both signs as the Constellations are squaring each other. It will take on the illness of the Planets and those Constellation which affects it.

Libra – 7th House Venus

The Libra Constellation is active and an air sign it acts in partnership, and shows a caring for their partner and what together they can do. When activated by a planet there would be a love in whatever they do. Keywords 'We can'. A plant that is coupled or shaped like a kidney i.e. a kidney bean or kidney meat are all in helping a healthy kidney. Needless to say it's anatomy rules the Kidneys and associated areas.

The 7th House is one of partnership. In illness it points to the Kidneys and Urinary Track.

The Planet Venus rules both Libra and Taurus. In Libra it rules love of partnership, it's active/cardinal and has an air element, which means it like to talk about the things they love. I have often combined the Taurus and Libra elements to come to a product i.e. sweet meat. Taurus is shown by the buffalo and rules meat, and I'd look for an alkaline product which is sweet to combine it.

Scorpio – 8th House - Pluto

The Scorpio Constellation is fixed and have deep concepts it rules real estate, regeneration, death, the responsibility of other people's money, and sex. It also rules the masses. It has the strongest faith when it decides upon one, possibly to the point of fanaticism. Scorpions will sting themselves under stress. Keyword 'I manage'. It is the poison of the Scorpio that would be useful in male and female reproductive system, or the end part of the bladder, the vagina and where it involves dead tissue. Its anatomy rules the Reproductive System, the Bowel, Bladder and Vagina.

The 8th House shows the groin and the Reproductive System or bowel would be affected if planets or constellation ruling it were afflicted.

Pluto brings with the influence of the lowest degradation and the highest of honors, the wealthiest of the wealthiest, but there again it will depend on what area it's found and the aspects surrounding it. It will add to a Constellation or House Calcium Sulphate; this can deaden or bring to life whatever the problem is but as you'll see in formulas I've created there is a great deal more involved. Pluto is a generational planet which means it will be in a Constellation for at least 13 years and they say as long as 32 years. By which time there are a generation of people born. It had made a spiritual aspect to Neptune for 15 years before I was born and if I'm right for 15 years + more. It totally tires a person out. You can sleep forever and watch life go by.

Sagittarius – 9th House - Jupiter

The Sagittarius Constellation is talkative and fiery it rules philosophy, sport, and higher education, overseas travel, countries and their people. it is hopeful in nature and speaks of ideologies. The Archer draws its Arrow and knows not where it lies (a philosophy is not necessarily practical), but a skilled archer knows where it lies. Its keyword is 'I philosophies'. In terms of illness it relates to heat and spreading as opposed to concentrated small areas, we could use the product of a horse to stop expansion and calm the liver. It also rules sand and silica. Its anatomy rules the Liver, Skin and Hips.

Something I've found interesting is that in early depictions of the symbol of Sagittarius it showed as a Centuarian a man with a horse's body. As such I've utilized that philosophy and have used a horse's manure and blended it with oil. After telling the volunteer client what I was using of which he agreed to, the symptoms he had went away.

The 9th House points to the Liver and Skin, hips/thighs gluttonous with Venus; if philosophies are being constantly challenged which is the external influence it can affect the internal actions of the liver and hips, and can lead to skin problems. The outside facilities of 9th house are universities, the seaside, sports fields, immigration, airports

Jupiter brings with it hope, optimism, the ability to renew one's faith, just as the Liver can renew and regrow itself. It rules the sands left by the oceans waves, clays with their elasticity and glass as a product of sand. It rules religion, law and philosophy

Capricorn – 10th House - Saturn

The Capricorn Constellation is an earth and active sign, seen by it's triplicities and Quadruplities; it rules pain, blockages, structure, and government it works towards structuring philosophy or ideologies of the previous sign causing it to become and create businesses. The Goat climbs the mountain to the highest peak. 'I work'. It can be humble always with a view in mind. As an earth product and refined we look to such products as oil, being cardinal, the sign would keep working on its environment, it rules coal, a product of coal is oil which is good for the knees and bones. Another form of Capricorn is bricks and cement. Its anatomy rules Knees and Bones.

10th house points to structure, and in the anatomy what supports the structure, but the knees and bones. It's cell salt is Calcium Phosphate and a product of that is potash. It is recommended that we take calcium for the bones, but not all calcium's are good for the bones which should be looked into. If Saturn wasn't in the 10th house it would mean it was likely that there'd be an absence of courts and government structure. If Saturn was in Taurus we'd probably see Goats as a food source, but, in Australia we have the Moon in Taurus where we are beef eaters and because we have an aspect to 10th house it is a commercialized item and transported overseas. Further investigation shows Venus in Sagittarius making an aspect to Saturn in Capricorn

The Planet Saturn brings with it a structured area; what a person would like to work on, it is serious and determined and rules courts and governments; it will bring with it Calcium Phosphate.

Aquarius – 11th House - Uranus

The Aquarius Constellation is fixed and an air sign, planets seen in this constellation show unpredictability or revolutionary ideas, it rules groups, animals, volcanoes, sudden eruptions. The Water Bearer nurtures. Keyword 'We think' We'd be looking for a food that would be unpredictable, such as mushrooms, this would probably help the circulation. It's anatomy rules the Ankles and Circulatory System.

A product in our oceans is created by the Volcanic reaction producing Chloride Ion mixing with Sodium Ion of the rivers, it would also be boiling underground water and mixing with the Chloride Ion, this creation is the tissue salt of Aquarius and Uranus and would be required by the 11th and 5th house. The basic element is Salt aka Sodium Chloride.

11th house rules groups, whatever Constellation is on this house will show what type of group. The 11th house points to the ankles and how a person circulates.

The Planet Uranus, is where a person or even a country will show their unpredictability and ingenious nature.

Pisces – 12th House - Neptune

The Pisces Constellation is talkative and rules the hidden area i.e. sleep, dreams, sleep talking, hypnosis, meditation, what we say without thinking or even realizing we've said it, it rules drugs and alcohol, ghosts, and, the supernatural, fears and phobias, illusions but also the psychic. In the external world institutes, prisons and mental health, mists, and, the knowing of something before it happens something that has no structure or form. The Fish going in different directions but returning to its group, an inheritance from Aquarius, is a possible reaction to being forewarned on a psychic level, predators of fish are often bigger fish. Its anatomy rules the Feet and Lymph System.

12th house points to the feet and lymph system. Whatever planets are in the 12th house will show the action that will take place on a spiritual or subconscious level. Areas it also points to are spiritual churches, Widgey Board aka Ouija boards, Séances. Environmental areas would be rivers, creeks, gullies where mist might form.

Neptune is the planet that rules the ocean along with Pluto whilst it's symbol is the Scorpion which can be found in many areas land and sea, and a new form with every constellation/house and planet. Neptune brings with it a spiritual or confused aspect or both; a person sometimes finds it hard to come to terms with forgetfulness, where they could have been daydreaming or falling asleep. The cell salt is Ferr Phos. And the quantity would depend on the condition of the illness. Personally I have many Neptune Pisces aspects and take 11 with a 100 in a container I can run out fast, where alcohol can be a substitute.

Chapter 7 - Tissue/Cell Salt Combinations

The Cell Salts of the Body and What They Mean by Earl Mindell

Below is an extract from 'The Vitamin Bible' - published in 1982 it's an outline of the Nutritionist Earl Mindell his properties of the natural chemicals found in the body, these are connected to various constellations, whilst there are number I'm only referring to 10. I've found in some cases Mindell's natural chemicals refer to the opposite sign properties or trines and sextiles. In the later part of this chapter I've also created a graph where I show the 2 natural chemicals that are combined by Dr. Schuessler and cover the 12 constellations, the graph shows how various combinations are being used for different parts of the body covered by the anatomy in the constellation. I've combined in the graph the supported evidence where hospitals are in some cases not all using the individual chemical as suggested by tissue salts in Astrology for their use.

In the mid 90's I used Tissue Salts frequently, but the dosage was highly increased when taking Ferr Phos, I was told I shouldn't have been walking in the late 90's with the problem with the white cell. count. At the time I was severely anaemic which I've gone into in later chapter and as such I was taking 11 Ferr Phos tablets; white cells are ruled by Pisces, and tiredness is one of Pisces problems. I wouldn't recommend it for other people unless you've been diagnosed as being extremely high or low in white blood cells requiring iron - Ferr is an iron whilst Phosphate is in abundance in our food and is an organic compound formed in a reaction been an acrid and an alcohol with the elimination of water Ref Bing Dictionary. I even went to the extent of drying my own seaweed because I couldn't buy it from health food stores when the Tissue Salts were taken off the shelves, but these were not the end all to the treatment I took. I was constantly being monitored by Dr's who'd read my blood tests results and when I asked them what I should take they'd say there was nothing on the market, as a result I'd ask if cell salts would help, they'd respond 'there was no medical research that contradicted them or supported them', so I was told, I could give them a try.

Another time I trialled Tissue Salts again in the late 90's, was when I had an infection in my finger and the fingertip was going black, I found when I analysed my emotions that I was beginning to panic which is an over concern, a worry out of control which I felt was Cancer - ruler of emotions. so as I saw the inflammation move I took about 4 Calcium Fluoride tablets, and to control it further, the color black made me consider Capricorn it's Tissue Salt - Calcium Phosphate which I took 1 of, when I realised it happened so suddenly I then added Aquarian Tissue Salt – Sodium Chloride tablet. The full experience resulted in my finger returning to a healthy color.'

When the tissue salts were stopped I began to look into the natural foods more for formula's and found it hard to find books that had a good joint analysis as such I've written out the individual properties below.

PROPERTIES OF INDIVIDUAL MINERALS

CALCIUM – Cancer, Scorpio and Capricorn

There is more calcium in the body than any other mineral.
Calcium and Iron are the two minerals most deficient in a woman's diet.

IN ADDITION
Maintains strong bones and healthy teeth.
Keeps your heart beating regularly.
Alleviates insomnia.
Helps metabolise your body's iron.
Aid your nervous system, especially in impulse transmission.

DEFICIENCIES
Rickets, hypoglycaemic, osteomalacia, osteoporosis. Growing pains and backaches could be helped with dolomite, chelate calcium or bonemeal.

BEST NATURAL FOODS
Milk, and milk products, all cheeses, soybeans, sardines, salmon, peanuts, walnuts, sunflower seeds, dried beans, green vegetables.

SUPPLEMENTS
Dolomite is a natural form of calcium and magnesium and no vitamin D is needed for assimilation.

ENEMIES
Large quantities of fat oxalic acid (found in chocolate and rhubarb) and phytic acid (found in grains) are capable of preventing proper calcium absorption.

ZODIAC SIGNS
Cancer – Calc Fluor (Calcium Fluoride); Scorpio – Calc Sulph (Calcium Sulphate); Capricorn – Calc Phos (Calcium Phosphate)

CHLORINE – Gemini and Aquarius
Chlorine regulates the blood's alkaline acid balance.
Aids in the cleaning of the body's waste by helping the liver to function.

IN ADDITTION
Aids in digestion
Help keep the body limber.

DEFICIENCES
Loss of hair and teeth.
BEST NATURAL FOODS
Table salt, kelp, olives.

ZODIAC SIGNS
Gemini – Kali Mur (Potassium Chloride); Aquarius – Nat Mur (Sodium Chloride – Salt)

FLUORIDE – Cancer
Fluoride is found in nature in the compound of calcium fluoride. It is sometimes added to water by the government it was believed it would improve teeth and lessen cavities

BEST NATURAL FOODS
Fluoridated drinking water, cranberry juice, seafood, blue crab canned and gelatine.

TOXICITY
Severe cases of fluorosis cause the enamel to become pitted and very badly stained. More a problem with children when their teeth are developing.

ZODIAC SIGN
Calcium Fluoride – Cancer

IRON - Pisces
Iron is essential and required for life, necessary for the production of haemoglobin, red pigment in muscles.
Iron and calcium are the two major dietary deficiencies of women.
Only about 8 percent of your total iron intake is absorbed and actually enters your bloodstream

IN ADDITION
Aids growth.
Promotes resistance to disease.
Prevents fatigue.
Cure and prevent iron-deficiency anaemia.
Brings back good skin tone.

BEST NATURAL FOODS

Pork liver beef kidney heart and liver, farina, raw clams, dried peaches, red meat, egg yolks, oysters, nuts, beans, asparagus molasses, oatmeal.

ENEMIES

Phosphoproteins in eggs and phytates in unleavened whole wheat reduce iron availability to body.

ZODIAC SIGN

Pisces – Ferr Phos (Phosphate of Iron)

MAGNESIUM - Leo

Magnesium is essential for effective nerve and muscle functioning.
Important for converting blood sugar into energy.
Known as the anti-stress mineral.
Alcoholics are usually deficient.
The human body contains approximately 21 g. of magnesium.

IN ADDITION

Aids in fighting depression.
Promotes a healthier cardiovascular system and helps prevent heart attacks.
Keeps teeth healthier.
Helps prevent calcium deposits, kidney and gallstones.
Brings relief from indigestion.

BEST NATURAL FOODS

Figs, lemons grapefruit yellow corn, almonds, nuts seeds dark-green vegetables, apples.

ENEMIES

Diuretics, alcohol

ZODIAC SIGN

Leo – Mag. Phos (Magnesium Phosphate)

PHOSPHORUS – Aries, Leo, Libra, Capricorn and Pisces

Phosphorus is present in every cell of the body. The most of it are in the above signs.
It's involved in virtually all physiological chemical reactions.

It is necessary for normal bone and tooth structure and Important for heart regularity and, essential for kidney functions.
Phosphorus is needed for the transference of nerve impulses.

IN ADDITION
Aids in growth and body repair
Provides energy and vigour by helping in the metabolisation of fats and starches.
Lessen the pain of arthritis.
Promotes healthy gums and teeth.

DEFICIENCIES
Rickets, pyorrhea – disease of the tissue that surrounds and support the teeth.

BEST NATURAL FOODS
Fish, poultry, meat, whole grains, **bonemeal,** eggs, nuts seeds.
ENEMIES
Too much iron, aluminium and magnesium can make phosphorus inactive.

ZODIAC SIGN
Aries – Kali Phos (Potassium Phosphate); Leo – Mag Phos (Magnesium Phosphate); Libra – Nat Phos (Phosphate of Soda); Capricorn – Calc Phos (Calcium Phosphate); Pisces – Ferr Phos (Phosphate of Iron).

POTASSIUM - Aries
Works with sodium to regulate the body's water balance and normalise heart rhythms. (Potassium works inside the cells; sodium works just outside them.)

In addition, it can assist in reducing blood pressure, edema, hypoglycaemia (Low sugar).

BEST NATURAL FOODS
Citrus fruits, watercress, all green leafy vegetables, mint leaves, sunflower seeds, bananas, potatoes.

TOXICITY
25 g. of potassium chloride can cause toxicity.

ENEMIES

Alcohol, coffee, sugar, diuretics

ZODIAC SIGN

Aries – Kali Phos (Potassium Phosphate); Gemini – Kali Mur (Potassium Chloride); Virgo – Kali Sulph (Potassium Sulphate)

SILICA - Sagittarius

Silica strengthens the blood vessels, improves elasticity in the joints. Prevents the body from absorbing aluminium and flushes aluminium from the tissues. Delays the tissues ageing process. Increases mobility. Aids in bone healing. Strengthens hair. Prevents wrinkles. Strengthen nails.

IN ADDITTION

Silica is required in walls of our body i.e. tendons, blood vessels, bones, arteries and separately the connective tissue. Aorta contains silica.

DEFICIENCES

Believed to cause Atherosclerosis.
Lack of elasticity in the skin.

BEST NATURAL FOODS

Oats (grounded), celery, yellow peppers, carrots, onions, unrefined grains, potatoes, cereals and white rice, yeast, beetroot.
The herb horsetail.

ZODIAC SIGN

Sagittarius - Silica

SODIUM – Taurus, Libra and Aquarius

Sodium and potassium are both found together and both found to be essential for normal growth. Sodium aids in keeping calcium and other minerals in the blood soluble.

IN ADDITTION

Sodium can aid in preventing heat prostration or sunstroke, and can help nerves and muscles to function properly.

DEFICIENCES

Impaired carbohydrates digestion, possibly neuralgia.

TOXICITY

Diets high in sodium usually accounts for high blood pressure.

BEST NATURAL FOODS

Salt, shellfish, carrots, beets, artichokes, dried beef, brains, kidney, bacon

ZODIAC SIGN

Taurus – Nat Sulph; Libra – Nat Phos (Sodium Phosphate); Aquarius – Nat Mur (Sodium Chloride)

SULPHUR – Taurus, Virgo, and Scorpio

Sulphur is essential for healthy hair, skin and nails.

Helps maintain oxygen balance necessary for proper brain function.

Works with B-complex vitamins for basic body metabolism, and is part of tissue-building amino acids.

Aids the liver in bile secretion

No RDA has been set, but a diet sufficient in protein will generally be sufficient in Sulphur.

IN ADDITION

Tones up skin and make hair more lustrous.

Helps fight bacterial infections.

DEFICIENCIES

Unhealthy looking hair.

Brittle hair.

BEST NATURAL FOODS

Lean beef, dried beans, fish, eggs, cabbage.

ZODIAC SIGN

Taurus – Nat Sulph (Sulphate of Soda); Libra – Nat Phos (Phosphate of Soda); Aquarius – Nat Mur (Chloride of Soda or Sodium Chloride)

Planet & Constellation	Tissue Salt by Schuessler
Sun/Leo	Magnesium Phosphate
Mercury/Gemini	Potassium Chloride /Kali Mur
Mercury/Virgo	Sulphate of Potash/Kali Sulph
Venus/Taurus	Sulphate of Soda/Nat Sulph

Venus/Libra	Phosphate of Soda/Nat Phos
Moon/Cancer	Fluoride of Lime/Calc Fluor
Mars/Aries	Potassium/Phosphate
Jupiter/Sagittarius	Silica
Saturn/Capricorn	Phosphate of Lime/Calc Phos
Uranus/Aquarius	Chloride of Soda/Nat Mur aka Sodium Chloride (table salt)
Neptune/Pisces	Phosphate of Lime/Ferr Phos
Pluto/Scorpio	Sulphate of Lime/Calc Sulph

CONSTELLATIONS CELL/TISSUE SALT COMPOSITION

Below is the basic cell structure believed to make up the human body formed by Schuessler in tissue remedies. Separately as indicated the medical field uses the advised tissue salts but, it needs to be realized that it is the combination of the tissue salts that work the best. In saying that any product used in excess can cause the same problem as seen in insufficient chemical composition, and by doing individual tests on the chemicals mentioned a doctor can restrict or increase as necessary. Separate research has been provided by http sites m.webmed.boots.com - Potassium; victoriahealth.com – Silica;

Areas	Kali	Phos	Mur	Nat	Sulp	Calc	Fluor	Silica	Mag	Ferr	Beneficial Uses
Aries /Mars /1 Skull/Head & Muscles	●	●									♈ Potassium Found to strengthen Muscles
Taurus/Venus/2 Neck/Thyroid & Growth Hormone				●	●						♉ Natrium is a Soda aka Sodium - helps Thyroid hormones.
Gemini/Mercury/3 Lungs Nervous & Respiratory System	●		●								♊ Mercury in too much quantity found to destroy brain cells. Suggestion - air mercury 0.02mg for mental problems
Cancer /Moon/4 Stomach & Digestive System						●	●				♋ Calcium used for teeth & whitening
Leo /Sun/5 Heart, Cardiovascular System		●							●		♌ Magnesium is used in hospitals for the heart
Virgo/Mercury/6 Intestines & Intestinal System	●				●						♍ Potassium is necessary for the intestinal wall & discharge, forces movement.

Libra /Venus/7 Kidneys, Venous System & Urinary Track		●		●							♎ Soda. In particular Baking Soda been found to be good for Kidneys
Scorpio /Pluto/8 Reproduction System					●	●					♏ Water Scorpion for reproductive area, large intestine & bowel.
Sagittarius/Jupiter/9 Skin, Liver & Integumentary System								●			♐ Silica for skin elasticity, hair, nails & liver.
Capricorn/Saturn/10 Bones/Knees & Skeletal System		●				●					♑ Calcium is known to help Bones
Aquarius/Uranus/ 11 Circulation & Ankles			●	●							♒ Chloride and Natrium make up salt needed for circulatory system
Pisces/Neptune/ 12 Feet/Lymph		●								●	♓ Ferrous is used for lymph problems which is pig iron

*

Chapter 8 - Gases

Not all planets have gases at least that is what they say, but even the hottest planets will omit some form of gas. Many curers have been brought about by single and combined gases, which I've found to be related to the type of illness related to the planet e.g. Helium is used for respiratory problems and helps to protect the brain tissue. Helium is found in Mercury the planet that rules the respiratory system and the brain

(Ref. Med Gas Res. 2013 Aug – www.ncbi.nlm.nih.gov/m/pubmed/23916029/)

Knowing the right amount for each patient would come down to a specialist in the field or even the specialist performing some experimenting especially for terminally ill people, unfortunately I haven't the facilities to look into this further.

Earth's atmosphere is made up *primarily* of only 2 elements - Nitrogen with 78% and oxygen of 21% as this makes up 99 percent it's obvious all other gases are very minimal.

Hydrogen is very combustible as a pure element mixed with oxygen is what makes water, they state Hydrogen is ruled by Saturn/Capricorn/10th house whilst I've noticed it in a number of planets and Jupiter with 90% in its atmosphere.

Sulphuric Acid is a product formed by the burning of coal and oil the natural product is Sulphur dioxide when omitted into the air it is evaporated by the sun and forms clouds mixed with moisture it's diluted into Sulphuric Acid also known as acid rain.

As we look into gases we find many gases are formed by the burning of metals/minerals and vegetation whatever it may be made up of, makes specific type of gas probably too many to mention.

Carbon Dioxide accounts for 0.03% of the earth's atmosphere and is ruled by Saturn/Capricorn/10th house strangely it takes up half the size of our body in veins.
Nitrogen makes up 78% of earth's atmosphere and is ruled by Uranus/Aquarius/11th house
Oxygen makes up 21% of the air is ruled by Sun/Leo/5th house
Helium ruled by Mercury/Gemini/3rd house
Ether in eyes and spinal canal ruled by Uranus/Aquarius/11th house. Ether is believed to be organic; but, it is also believed to be the outer area of our atmosphere.

Below is a chart of the atmosphere that are surrounding each planet. These could change depending on what literature you read. It could also change depending on the influence of one planet on another when making any type of aspect. This could alter the accuracy of literature we read.

I have found it quite amazing how Carbon Dioxide with such a low percentage has such a powerful impact on earth. 'The atmosphere has increased by more than 40% since the start of the Industrial Revolution, from 280 ppm in the mid-18th century to 402 ppm as of 2016. The present concentration is the highest in at least the past 800,000 years'. (https://en.wikipedia.org/wiki/Carbon_dioxide_in_Earth%27s_atmosphere). One milligram in a kg is 1 ppm (by mass). One liter (L) of pure water at 4ºC and 1 standard atmosphere pressure weighs exactly 1 kg, so 1 mg/L is 1 ppm.

Carbon Dioxides is a Saturn influence and under the Encyclopaedia of Medical Astrology states it's a dominant element in earth signs Taurus, Virgo and Capricorn. I'm stating this because it doesn't show up under the cell salts; It causes constraint and retention and Saturn is the main influence in heart attacks. Saturn is known for being a building block building one cell after another.

Planets where Gases and Minerals are found.

	Sun	Mercury	Venus	Earth	Mars	Jupiter	Saturn	Uranus	Neptune	Pluto
Argon		✸	✸	0.93%	1.60%					
Calcium		✸								
Carbon Dioxide		✸	96.50%	0.04%	95.32%					✸
Carbon Monoxide					0.08%					✸
Helium	7.80%	6%				10%		15.20%	19%	
Hydrogen	92.10%	22%			✸	90%	✸	82.50%	80%	
Hydrogen Deurteride									192 pt/ million	
Krypton		✸			✸					
Magnesium		✸								
Methane						✸		2.30%	✸	✸
Neon										
Nitrogen		✸	✸	78%	2.70%		✸			✸
Oxygen		42%		21%	0.13%		✸			
Potassium		0.50%								
Radon										

Sodium		25%								
Sulphur/ Sulphuric Acid			✸			✸				
Water/ Vapour		✸	✸	0.4%- 1% at Sea	✸	✸		✸		
Xenon		✸			✸					

Ref: www.space.com

Chapter 9 - Working Within the Hours of the Day

This may sound simple but when you compare your own hourly chart you may find yourself working against the celestial hourly chart. This is what I believed happened to me when working on NEVILLE R..'s chart. As soon as I brought something to his attention he wanted me to try it and when I brought to him the hourly system he wanted me to work by it, I did this when doing acupuncture, hypnosis and massage. This I feel you should remember because without a proper analysis it could be psychically draining. What that meant to me was I was totally exhausted and never wanted to do it again.

Before looking at the planets and the hour they rule let's look at the horoscope, because there are planets that are in their detriment and others that are exalted and so if there is a particular planet that is ruling a particular hour it could be working against you or your client's horoscope if so it could be draining more than instructive. But, who am I to speak, I do everything at the wrong time and I'm still here and my friends have seemed to benefit.☺

Planets in their Detriment, Fall and Exaltation

A planet is said to be in its detriment when it is opposite the sign of its rulership. This causes a planet to be limited in its expression of its basic characteristic.

A planet is said to be in its fall when it's opposite to the sign it is exalted in. When a planet is in its fall it is said to be more hazardous to health. It loses its strength and influence.

Planets are strongest in those signs in which they are in their exaltation or rulership. This means they can clearly express the rulership of the planets in that sign. It is good to remember the area of health each planet and constellation rule. For instance, signs in their exaltation i.e. Sun in Aries give thought to the heart and red blood. Moon in Taurus gives thought to stomach/digestive track and throat/thyroid and hormone development. As we continue Mercury in Aquarius gives thought to Brain/Arms/Oxygen/Air and Circulation/Ankles/Electro neurons. Venus in Pisces give thought to Veins/Kidneys/Loin and Lymph system/Feet/clear white cells and antibodies of the body that fight bacteria. Mars in Capricorn gives thought to Red Blood Cells/Blood/Muscles and Bones/Knees. Jupiter in Cancer gives thought to Liver/Blood purifier and Stomach/Digestive track. Saturn in Libra gives thought to Knees/Bones/Vertebrae and Kidneys/Veins. Uranus in Scorpio gives thought to Circulatory system/carrying the oxygen in the blood/ankles and Reproductive Organs/Groin/Hormonal changes. Neptune in Cancer gives thought to Mists/lymph/feet and stomach/digestive track both are water signs so this also adds to the thoughts. As we go on we can extend it to pages if we thought about what each planet and constellation means. I'm just giving the basic structure which if studied individually is very broad. The last in the exaltation is Pluto in Leo, which is the Reproduction Organs/Groin/Gonads and Heart/Blood Circulation.

Detriment	Fall	Exalted
Sun in Aquarius	Sun in Libra	Sun in Aries
Moon in Capricorn	Moon in Scorpio	Moon in Taurus
Mercury in Sagittarius	Mercury in Leo	Mercury in Aquarius
Venus in Scorpio	Venus in Virgo	Venus in Pisces
Mars in Libra	Mars in Cancer	Mars in Capricorn
Jupiter in Gemini	Jupiter in Capricorn	Jupiter in Cancer
Saturn in Cancer	Saturn in Aries	Saturn in Libra
Uranus in Leo	Uranus in Taurus	Uranus in Scorpio
Neptune in Virgo	Neptune in Capricorn	Neptune in Cancer
Pluto in Taurus	Pluto in Aquarius	Pluto in Leo

Now let's look at the Planets and the Hours they rule

First it all begins with a date - 31.7.2001 – the Zodiac sign is Leo, changeover always takes place between 21/22 of each month and lasts till 21/22, an Ephemeris will give you the exact sign if born on the cusp; more importantly for what we are looking at is the exact time the Sun rises and sets; I've always associated Leo with the colour bright yellow but some may say Royal Blue. The zodiac is a proud and noble sign and so people will be more proud and noble than usual throughout the Leo month.

The Day the Planets Rule

The planets flow as follows for week days: -
Sun, Moon, Mars, Mercury, Jupiter, Venus, Saturn and this is fairly easily to remember as the Sun is Sunday, Moon is Monday, Saturn is Saturday and they are 3 names already given you by knowing the planets.

Tuesday and other days, we need to know a bit of Latin/Germanic or the Nordic mythology as Tiw means Mars and Tuesday came from Tiw/Tyr – Norse god of war, it was Tyrsdag finally it became Tuesday, as such Tuesday is an action day and a good day to begin things.

Wednesday was named after Wodan his role was Father of all Gods, also known as Oden. From Wodan came Wednesday. He had to do with wisdom/ poetry, war and death. He was able to speak to the dead to question the wisest, so it's a communicative day and one of deep thought. It's also ruled by Mercury = Mercurii. God of trade/profit/merchants and travellers, the original trade was in corn.

Thursday was named after Thor god of thunder, one of the most powerful gods. He was said to be protector of gods and humans Thor became Thursday.

Friday was the day named after goddess Frigg wife of Oden/Wodan, patron of marriage/motherhood/love and fertility. This is maybe why we often hear people say I love Fridays.
Ref. Micha F. Lindemans – Encyclopedia Mythica

The Hour the Planets Rule with Separate Association to the Day

The hours have a different rulership to the planets that rule the days of the week. It must be remembered this alters for the hours of 24-hour period. It flows Sun, Venus, Mercury, Moon, Saturn, Jupiter, Mars

The days of the week the planets rule are:
Sunday – Sun
Monday – Moon
Tuesday – Mars
Wednesday – Mercury
Thursday – Jupiter
Friday – Venus
Saturday – Saturn

The first hour after Sunrise on a particular day is ruled by its sign of that day, so if it is a Sunday, the Sun will rule the 1st hour after sunrise. The Moon will rule the first hour from sunrise on Monday etc.

The span that is ruled by the planets is found by calculating the time from sunrise to sunset, which alters from month to month, from season to season from zone to zone and don't forget times are adjusted for daylight saving. In this case Sunrise is 5:40 a.m. Sunset was 6:30 p.m. This gives us 12 hours 50 minutes for the whole daylight period on the 31.7.2001, and, so it's not necessarily going to be an even hour. We then divide 12 h 50 m by 12, as 12 hrs rules night and day and equals 24. When we divide 12 h 50 m by 12 it equals 1 hour 4 minutes with 2 minutes over. We'll take it to the nearest minute which will leave 2 minutes over which is why the last time is 18:28 p.m. These are daylight hours, if you want to be more precise you can, it will take it into seconds which would be 5.40.10 a.m. but I'm rounding it off.

We'll begin our week with a Sunday and divide the daylight hours by 12 and therefore each planet will occupy 1 hour and 4 minutes of day light hours.

Please note the change in rulership of hours to those of days.

For Sunday

1st hr ruled by the Sun	5:40 – 6:44 a.m.
2nd hr ruled by Venus	6:44 – 7:48 a.m.
3rd hr ruled by Mercury	7:48 – 8:52 a.m.
4th hr ruled by the Moon	8:52 - 9:56 a.m.
5th hr ruled by Saturn	9:56 – 11:00 a.m.

6th hr ruled by Jupiter 11:00 – 12:04 p.m.
7th hr ruled by Mars 12:04 – 13:08 p.m.
8th hr ruled by Sun 13:08 – 14:12 p.m.
9th hr ruled by Venus 14:12 – 15:16 p.m.
10th hr ruled by Mercury 15:16 – 16:20 p.m.
11th hr ruled by the Moon 16:20 – 17:24 p.m.
12th hr ruled by Saturn 17:24 – 18:28 p.m. Sunset

The nightfall or sunset to sunrise is calculated the same way, sunset I've rounded off to 06:30 p.m. or 18:30 hrs, sunrise is 5:40 a.m. the next day, the interval then is 11 hrs 10 mins, multiply 11 x 60 = 660 minutes + 10 minutes = 670 divide this by 12 this comes to 55.8333 minutes. I've brought it to 56 which puts it out by 2 minutes; 2 minutes multiplied by 60 = 120 seconds / by 12 = -5 seconds for each hour if you want to be precise.

The pattern on the role of planets has already been set by the above calculations for sunrise to sunset, we then continue with the next planet after Saturn ♄ which ruled the last 12th.

As such it would follow for nightfall 6:30 p.m:-

For Sunday
1st hr is ruled by Jupiter 18:30 – 19:26 p.m.
2nd hr is ruled by Mars 19:26 – 20:22 p.m.
3rd hr is ruled by Sun 20:22 – 21:18 p.m.
4th hr is ruled by Venus 21:18– 22:14 p.m.
5th hr is ruled by Mercury 22:14 – 23:10 p.m.
For Sunday / Monday
6th hr is ruled by Moon 23:10 – 00:06 a.m.
7th hr is ruled by Saturn 00:06 – 01:02 a.m.
8th hr is ruled by Jupiter 01:02 – 01:58 a.m.
9th hr is ruled by Mars 01:58 – 02:54 a.m.
10th hr is ruled by Sun 02:54 – 03:50 a.m.
11th hr is ruled by Venus 03:50 – 04:46 a.m.
12th hr is ruled by Mercury 04:46 – 05:42 a.m

An adjustment for the next day begins at Sunrise, if it's Tuesday Mars begins the first hour at sunrise etc.

By this section you should know the planets meaning; and that this book is a reference book for astrologers. I'm saying that because I'm reflecting on the characteristic of the Sun, when it's not afflicted it brings clarity and it rules the heart, some say it's the expression of the will. If it's afflicted by Mars there can be too much heat especially if Mars is in a fixed sign, remember that Mars rules the red blood cells and the muscles, if it is in a fixed sign of Aquarius which is opposite Leo this is also fixed by it's quadruplicity, you have to remember to look at the house if the Sun is in 7th house we are looking at the person being active. This is where it becomes complex. 7th house is ruled by Libra its quadruplicity is cardinal. Mars in Aquarius would mean it would be stopping the fluids that is pumped by the electrodes in the circulatory system. This is because too much heat with Mars and Sun both fire sign would dry the air or oxygen up in the circulatory system, stopping the electrical pump in the system.

Chapter 10 - Colours

Colours are an important part in finding a cure, and I believe further studies are needed. Satellite images display colours that seem simple; but, they can be distorted by gases, a planet's surface may be red as in Venus whilst with the white clouds it becomes a brown/pink. The Earth appears blue and white because it is believed the water is evaporated into the atmosphere and the white clouds are due to the Sun's action, it doesn't mention the green we presume would be their due to forests and replanting of trees or the brown from the denuding of trees and the green the ocean normally is.

When I look for a formula an important procedure I do is to mix the colours of the constellations and planets and combine the conjuncts, those where another planet is at the same degree or near, I try and imagine what the colour then would be.

In illness there are many areas of discharge, just in the head we have ears, eyes, nose and mouth. In the reproductive area there are the bladder/urinary track, bowel/anus, vagina or penis for men. Whilst men use the same structure for urinary track and sperm, they are two separate structures within one. We also have pores in out body that discharge sweat, we don't think of it having colour but I have seen yellow fluid being released on a body that has just died. There are many different types of infection that changes the colour of the excreted fluids whether it be from your head/reproductive organ or pores it is the colour we are relating to here which we can associate with planets/constellations and even houses.

Colours of planets from satellite images

The old Astrologers who were all Astronomers were thought to not have modern day technology and so the colours they put to planets and constellations were assumed to be from their native eye, observed from earth, but, archaeologists are finding unbelievable instruments in tombs and elsewhere that's hard to explain and in latter times they've found similarities to what was thought to be new inventions to those buried in tombs. I've found conflict in some Astrology site; one in particular is their colour of Libra being green, satellite imagery doesn't agree, as below, and I've found nothing to support it, except the gall bladder and oil it holds is green, as such I believe there are secondary colours.

The Satellite Colours

Mars	red-dark orange caused by rustic rocks.
Mercury	grey, it has no atmosphere and shows the rock surface, but it could also be the colour of mercury.
Venus	pinky brown - the atmosphere is yellowish-white which makes the surface rock red appear pinky brown which was found by the NASA the Magellan project.
Earth	blue, it is the atmosphere caused by the oceans and light where it also can be seen as light blue and white clouds.
Jupiter	orange and white bands, its believed the bands are coloured by ammonia clouds, and the orange comes from ammonium hydro Sulphideide clouds of which the planet is made up.

Saturn pale yellow, this is caused by white ammonia haze. In the planets winter the northern hemisphere appears blue.
Uranus light blue, caused by the methane clouds.
Neptune light blue (yet darker then Uranus), caused by methane clouds.
Pluto light brown, believed to have dirty methane ice on the surface.

V

The colours of planets I use to depict planets causing illness.

Some colours could be the association of other planets.
Mars red-orange
Aries as above
Venus pale pink, the colour of the venous system mixed with CO^2 - as associated to the Earth Element of Taurus.
Taurus reddish brown, a dirty red-maroon
Mercury Mercury's colours change to what planets are affecting it, as such it could be gray contributed to a cloud affect, also the name indicates a deadened gray or silver metallic colour.
Gemini as above
Moon white
Cancer white and black, indication of full moon and new moon.
Sun yellow-
red-mustard
Leo as
above
Mercury as seen as an earth's colour, mercury silver or deadened gray.
Virgo as above
Venus pink
Libra as
above
Pluto dark purple-dark brown-light maroon, black
Scorpio as above
Saturn dark brown/black
Capricorn brownie black
Uranus blue, the colour of the sky
Aquarius as above
Neptune Mist colour, green of lakes
Pisces as above

The below show the three initial colours called the primary colour which all other colours come from. Compliments of Google. I'm endeavouring to show how when each colour is mixed with another it can be understood by an astrologer, it is a specialized field so not all would have studied it, in saying that I feel my knowledge is limited, but, there is a lot I have learnt when making formulas and looking for specific colours. Heat changes a colour, water causes colours to be more translucent as do gases, colours mixed with earth gives them a degree of quality.

In astrology dark yellow is the colour of the Sun and ruled by the Sun, Leo and 5th house; blue is the colour of the Sky ruled by Uranus, Aquarius and 11th house. Pink is the colour of Venus, whilst the planet is a red rock the natural gases makes it look pink. If we look at our environment it's the atmosphere which relates to the gases that are transformed by the Sun causing photosynthesis which has brought about the greenery of our forests; where green is seen as a secondary colour.

The colours below are through darkroom filters, I found these interesting because they show black and white, Magenta has been created by red and blue, Cyan has been created by green and blue.

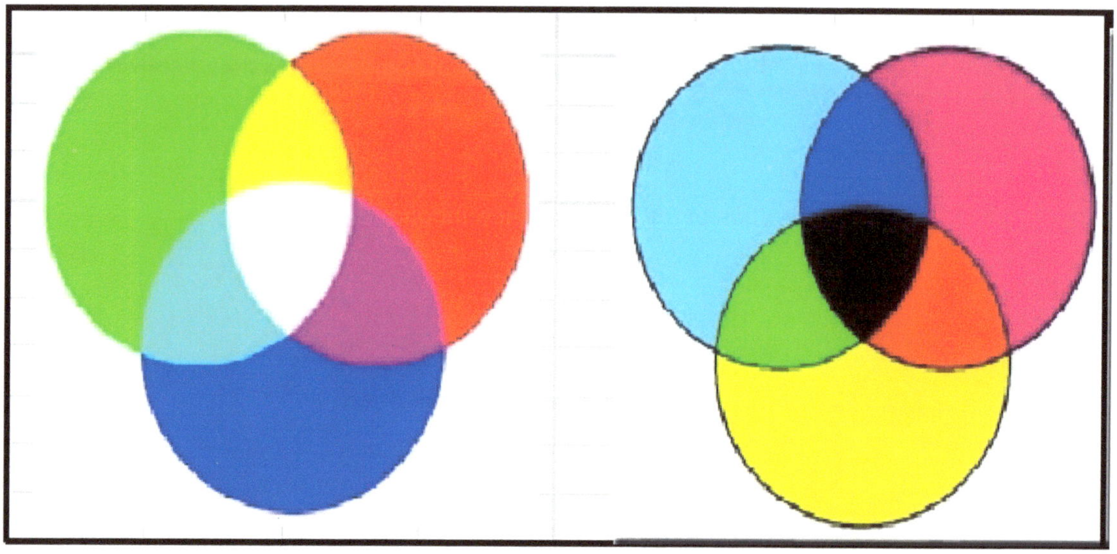

There seem to be a confusion to what the primary colours are, with the transparencies above we see how several colours affect the pure colours if there are such things as pure colours, as primary colours are colours that can't be made by other colours but I would say 4 colours red, blue, green and yellow seem to make more sense to me. This coincides with the findings of Ewald Hering. In astrology the opposites are necessary to make the squaring colour and there isn't a colour that can't be made it's just finding the right mixture.

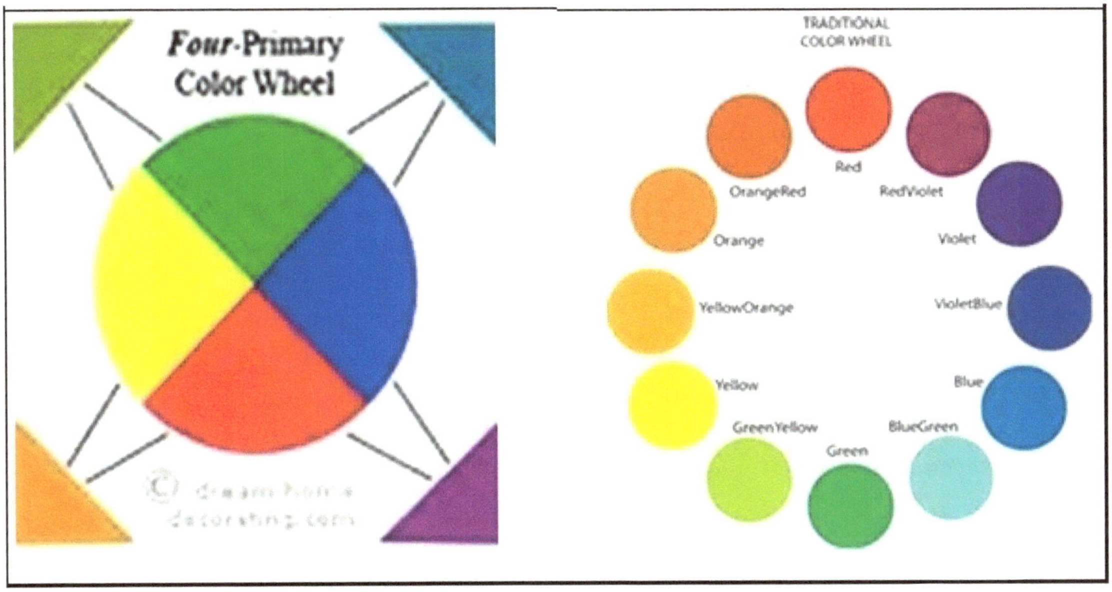

The above is an example of the four primary and colours made from those, found at *http://www.bing.com/images/search?q=images+for+free+of+4+Primary+Colours&id=15834B77D751AE749A400CB8B8F6BB7602B9095D&FORM=IQFRBA#a* Primary colours could be more looked upon in astrology as fixed colours.

Water Colours and How They Give a Wishy Washy Affect

The graph above is chart by James Gurney who painted a chart of watercolour pigment combinations. http://gurneyjourney.blogspot.com/2010/10/overlaid-wash-test.html

I've thrown in the water colours because there are 3 water signs/groups and if you're not using a pure chalk like pigment than there is some form of fluid in the paint which adjusts the colour, alcoholic bases are formed by Neptune/Pisces/12th house the other two are Moon/Cancer/4th house and Pluto/Scorpio/8th house these signs rule ocean/lakes/rivers/marshes and mists. Having mentioned Pluto, it made me remember that some painters used to use their own faeces or that of animals, which is manure at least this is what I saw in a documentary and if you think about a painter in the bush they'd need to be inventive to find their colours a few hundred years ago, but I understand it tendered to crack unless there was oil mixed with it. Personally, it wouldn't be a picture I'd seek to buy.

I love these Colours as they are really giving you the watery effect, that can only be seen with the water signs and their rulers. The changes are particularly noticed with the yellow, and how you can see how it changes the pinks and blues. Of course, I relate yellow to the Sun, Blues to the Earth but the paler shades the Sky – Aquarius.

In health a washed out look is often the look which leukemia gives can be contributed to Neptunian problem. An indigo/mauveish. wishy-washy look can often be contributed to Pluto an affect mixed with a paler colour maybe the white of the Moon in such a case one could look to the hormonal structure and stomach problems. I've noticed mottled colours are found in mutable signs i.e. Virgo, Pisces, Gemini and Sagittarius. Mottle colours are those with specks of colour or waves i.e. in clouds. We also see it a lot in horse races.

THE GIRL WITH ACNE

With colours and health problems colours are changed by the influence of other planets, which I've tried to show with the combinations, but if there was a pure colour our body actually tells us by what we see. To explain this, I once treated a girl who came to stay a few days, she had pimples on her face that were unusual, no one had been able to help her and nothing she bought from the supermarket had worked, I began telling her what constellation ruled the head, which is ruled by Aries, it ruled problems with pimples, which are spear shaped and red, its character is to be aggressive, active, but this had to be set off by something. I then explained to her what the cardinal signs meant which involved partnership, family, and constraint it eventually came out that her partner's family, his father in particular was going to be released from prison and had been a child molesterer. Her partner wanted his father to be part of his family, and only 6 months earlier she had given birth to a little boy. She had tried to fight her partner on this, but he said he couldn't do that to his father.

She had a huge decision to make. As she spoke her face was healing. The pimple colour changed so I had to change the treatment to incorporate the new colours. I can only remember the last treatment I gave her, was a pre-prepared food, actually supermarket bought, that combined the colours of blue and white and had been fermented to achieve this, which I asked her to put on the pimples. I looked for a food that had the same action the stomach does, for it has hydrochloric acid in the stomach which ferments its food, this action is ruled by Cancer. The action I took helped her to clear up what I now believe was a type of psoriasis but it wasn't scaly so there could be another name, which I can't find. I put it down to bringing up concerns she hadn't been able to speak about and the treatments I was able to do over the 3 days. When she left she was a much a happier girl and I believe that was due to being able to understand her situation. In my case when the colour of the symptoms changed it meant there was more research to do. By the third night she was not embarrassed to go out. :) The reference book I would have used would have been Encyclopaedia of Medical Astrology by H.L. Cornell M.D. and this is why I've expanded on colours below.

When there are problems and when these colours are seen you can look to these: -
Red is the colour of Mars - health signs: - redness, red skin, rashes, blood hyperactivity, headaches, sudden cuts, anything that appeared spear shaped. The colour relates to problems with actions.
Pink to dark crimson is the colour of Venus or Libra (not Taurus), problems with acid in urine, salt in urine, kidneys and urinary problems, veins, lower back pain. The colour relates to problems with love ones, if Taurus money.
Black is the colour of Saturn/Capricorn: - causes problems with blockages, crystallization, knee pain, black discolouration of the skin shows a blockage in the air flow. The colour relates to problems with government, work, and the older generation.
Yellow to orange is colour the Sun/Leo: - orange discolouration, means there are problems with the heart or oxygen in the blood. The colour relates to problems with entertainment, self and patriarch or son's.
Yellow to light yellow is the colour of Jupiter/Sagittarius and means problems with the liver. The colour relates to problems with philosophies, universities, religion all areas of an overseas nature.

If we complicate matters as below we see what happens with a mixture of colours: -
Magenta/purple is the colour of Venus mixed with the Sun /Leo comes red Aries
Cyan is the colour of Pisces mixed with Magenta Libra makes blue the colour of blue/Aquarius
Cyan Magenta and Yellow are the colours of Pisces/Libra/Leo or Neptune/Venus/Sun = Black
White is the colour of the Moon/Cancer and mixed with Neptune, which makes colours more penetrable, this could be seen in anaemia.

As such we can look at illnesses through colour. If a person has gone blue, they've been deprived of oxygen. Oxygen is pumped to the heart by the lungs it allows our body to puff up, it's an invisible gas and part of the air we consume which contains hydrogen, nitrogen, carbon dioxide. It's a sign that the heart has stopped working. The hearts function is to combine oxygen with blood, the blue veins have carbon dioxide in them which we can see if we put a tourniquet around our upper arm and watch the blue vein come to the surface and sometimes we only need to open our hand and look at our wrist, because it's not a clear colour we'd call it polluted and yet some parts of the body requires carbon dioxide which is brought to them through the mixture of blood and so it's been found that the colour comes from the mixture of carbon dioxide and blood. Ultimately they go to the heart to be cleansed and the arteries take the renewed blood, the red blood around our body, in fact the body has a recycling system. If we look at charts and figurines it looks like it's defined, where red takes over from blue but in fact it isn't. When the body dies the pores release all types of different colours which has been dependent on what we eat. But, there again colours could differ depending on what the illness has taken and what it has left.

Colour Chart

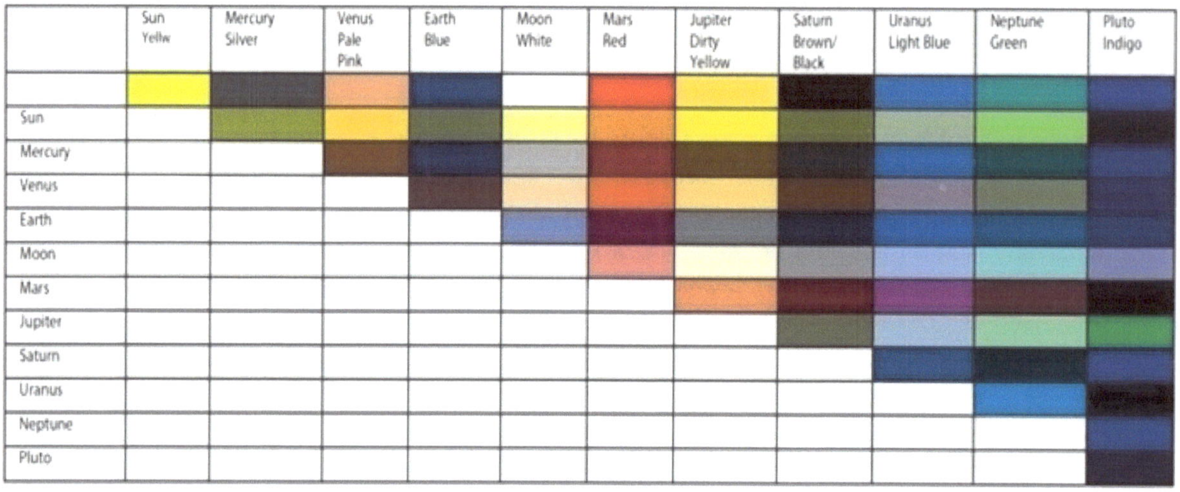

Sun Yellw
Mercury Silver
Venus Pale Pink
Earth Blue
Moon White
Mars Red
Jupiter Dirty Yellow
Saturn Brown/ Black
Uranus Light Blue
Neptune Green
Pluto Indigo
Sun
Mercury
Venus
Earth
Moon
Mars
Jupiter
Saturn
Uranus
Neptune
Pluto

<u>Illustrations of Colour of Planets when Conjunct</u>

I have created the boxes below using the chart above and adding the colours I've created to show a multi-colour affect, the centre colour is the outcome from it, it's the colour we can expect to see when there is planetary problems causing illness.

Yellow & Black /Sun & Saturn

When we look at the Sun it's generally bright yellow unless other factors have changed it. If it's fire which has turned the sky red, we're looking at Mars/Jupiter & Sun. Here we're looking at Saturn. Saturn blocks oxygen which is ruled by the Sun and without oxygen we die. Going a yellow grey is a colour of lack of oxygen, blockages to the heart are usually a build-up of calcification, these are calcium blocks/cells built on top of each other, the building of something is a Saturn influence but, calcium is part of its cell salt.

Pink & Yellow - Sun & Venus

Venus's colour is that of red rock covered by a cloud which makes it pink/brown. It is the colour for both Taurus and Libra. Whilst veins are ruled by Libra under the Venous System the colour of varicose veins is an actual adulteration of the natural colour of Venus and shows the lack of oxygen, but If we have the Sun and Venus together we are likely to see a pale orange. You could find a heated thyroid or heated kidneys if there are problems. Taurus and Libra have a natural health problem aspect if there are planets in those signs, but usual this isn't the case, only one other are shared by the same planet.

Blue & Yellow - Sun & Earth

The Earth has been called the water planet as is illustrated by blue, and whilst the Sun is always in the opposite sign to the Earth if there are additional problems and the fluids of the body are being affected we get the grey/yellow colour, much like that of varicose veins. A lighter mix of yellow and blue stands for circulation problem under which would be seen with Uranus.

Grey & Yellow - Greeny Blue - Jupiter/Mercury - Neptune

Here the colours become more complicated Jupiter and Mercury is the colour of Grey and Yellow mixed this gives the combined colour as seen in the Colour Table above when it's added to Neptune it becomes a smokey greeny/blue/brown which is the centre colour. Neptune does have a smokey/misty affect.

Illustrations of the Anatomy and the Zodiac

ANATOMY AND THE ZODIAC

This shows the relationship between the Zodiac and the Anatomy and the special relationship between the signs and the body. Often illness comes from polarity of the opposite sign, e.g. Aries – Libra, if there is a headache it could be related to a kidney problem ruled by Libra.

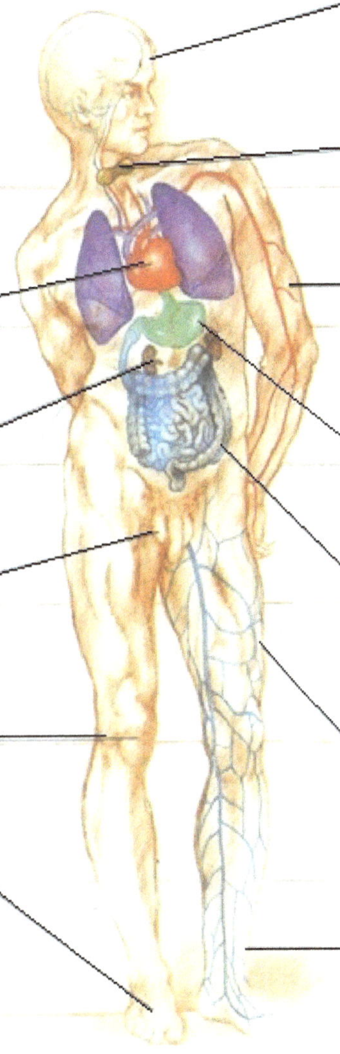

Leo – Aquarius
Leo rules the heart, spine & back. They have a tendency to enjoy themselves too much, they need to slow down as they get older for heart problems

Libra – Aries
Libra rules the kidneys any disruption to their lives can result in a back pain relating to the kidneys.

Scorpio – Taurus
Scorpio rules the sexual organs. They are more highly sexed then other signs. Any frustration or suppression can lead to unpleasant behaviour.

Capricorn – Cancer
Capricorn rules the knees, bones and teeth. Orthopaedic and dental troubles are common as with deterioration of the bones.

Pisces – Virgo
Pisces rules the Lymph system and the feet. As such they should always wear well fitted shoes, with a common ailment being anaemia.

Aries - Libra
Aries rules the head, they often suffer headaches. The glands are the subrenals, beneath the kidneys, which pumps adrenalin into the blood for emergencies

Taurus – Scorpio
Taurus rules the throat and neck. They are particularly vulnerable to colds and chills. The gland is the Thyroid any problems will lead to weight problems

Gemini – Sagittarius
Gemini rules the nerves arms and shoulders. Their prone to break collar-bones and arms. Internally lungs where colds often turn into bronchitis. Gemini's are restless live on their nerves.

Cancer - Capricorn
Cancer rules the stomach, mammary glands and alimentary canal. Their tendency is to get upset and worry more than most. They're prone to indigestion and ulcers.

Virgo – Pisces
Virgo rules the nervous system and intestines; they suffer from bowel & stomach problems. They can be worriers due to nervous tension.

Sagittarius – Gemini
Sagittarius rule the liver, hips and thighs, Sagittarius stagnate if they don't give themselves considerable amount of mental and physical exercise resulting in excessive weight.

Aquarius – Leo
Aquarius rules circulation and ankles. They can suffer from varicose veins and hardening of the arteries. Fractures are common in the shin and ankles

ANATOMY AND THE PLANETS

This gives a brief outline of the planets relationship over the endocrine glands that release hormones into the blood.

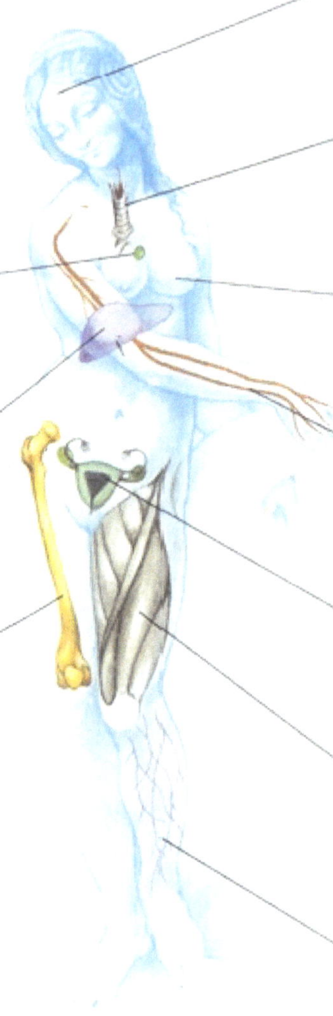

The Sun rules the heart, back and spinal column, and assoc. with the thymus, an endocrine gland behind the upper end of the sternum that is important to puberty. With a connection between the immunization against bacteria.

Jupiter
Jupiter has always been linked with the biggest gland, the liver, and it's purifying function. We now find a strong relationship to the pituitary gland. It regulates hormone production & governs physical growth.

Saturn
Saturn is connected with the gall-bladder, spleen, skin, teeth and bones. This later effect is strengthened by its influence on the anterior lobe of pituitary, which relates to the sex glands, bone and muscle structure

Mercury
Mercury is associated with the respiration, the brain and the nervous system which gives strong links to all parts of the body.

Venus
Venus rules the throat, kidney, lumbar and parathyroid. These play an important part in controlling the calcium levels in the body.

The Moon
The Moon is connected with the breasts and the whole alimentary, or food system - the oesophagus, stomach, liver, gall bladder, bile ducts, pancreas and intestines.

Uranus
Uranus is connected to the circulatory system, the gonads and the pineal body – an area of the brain important to primeval animals, but disputed by man. It's sometime called the 'third eye'.

Pluto
Pluto is linked with Mars in influencing the gonads. It's special connection is with the formation of cells and the reproduction function itself.

Mars
Mars acts upon the muscular and urogenital system and upon the gonads or sex glands – ovaries or testes.

Neptune
Neptune acts upon the general nervous system and in particular the Thalamus – a structure in the brain that gives a vital part in the transmission of stimuli to and from the sensory organs.

Chapter 11 - Example of How Charts Give You The Avenue & Procedure for Formulas

I've given formula's in the chapter below, and I've repeated the same type of graph for each person. Once you understand the meaning of the planets, constellation and houses you go to doing a graph for each of them, taking particular notice to the areas that have close degrees, with blockages usually with planets opposite, square and I find quincunx's causing an underlining problem. When I've had blockages I'd keep a flow chart up on my whiteboard for days before coming to a formula, if I could come to a formula, I know if someone was on the same page as myself it wouldn't have happened as I have had occasion to speak with one person where we did work on a Leukemia formula unfortunately nothing came of it.

For myself I could only explain it as writer's block. The thing is not all illnesses need to be treated with a formula only if they've been left untreated for too long. It's where Quadruplicities and Triplicities come in, which in the past I feel I've only just skimmed over. These are very important areas to a formula and you'll see in Mark's formulas I've put in a flow chart just for Quadruplicities and Triplicity alone just to show the Constellations Houses and Planets. I've had to do a separate section for aspects they make.

When we are looking at Quadruplities we only see 3 influences Fixed, Mutable and Cardinal meaning Stubborn for the lack of another word, Talkative and Movement. In the early stages in the mid 90's when I was trying to put things together I had a young girl say to me can you help me try and fall pregnant, I think she had not long miscarried and Doctors couldn't give her any hope; and out of the blue I said 'you have a number of fixed planets it would be best to stay still after sex also to elevate your backside'. I remember also saying she had to lay down to keep the babies'; this I thought was an impossible task but I heard nothing more until I found out a year later she had a baby and not long after a mother of two. :0)

With the 'mutable' signs I remember stressing with Wayne R. that the planets along with having a skin cancer, showed there was conflict as to their and his belief where he was living; and in a sense they were strangling him, not literally. The skin I've always considered to be under Jupiter/Sagittarius & 9th house rules philosophies and as such are Mutable/talkative, another symptom was that the skin cancer was all up his arms this brings in Sagittarius's opposite Gemini or its components being square is also Mutable, this implies it also has to do with communicating where it needed to be explained there was a problem with expressing himself. Another point away from Mutable was that the skin was crusty and I instantly thought a problem with Saturn; Saturn comes under Cardinal/moveable, and occasionally I see him scratching off the flaky skin or more wiping over the skin cancers with his hand. For Wayne I've written up his formula in the next chapter. Where in his flow chart the first thing you can see under 'Aspects' a square between Jupiter and the Sun/Self, next is Neptune a mutable sign square Moon, Mercury square Saturn and Saturn is conjuncting Jupiter both in Capricorn. What I haven't marked is an ill placed Constellation Sagittarius in 3rd house ownership of Gemini. These two constellations are normally opposite. The 3rd house and Sagittarius are both Mutable. There is also Mercury in Sagittarius, Mercury's the ruler of Gemini and also has a mutable element.

The main thing is for a patient who have problems with the areas that are Mutable to be able to talk about the areas planets are in it can take the tension away from the area that is being blocked. The same goes for Cardinal influences these are active signs and for instance it's recently been publicized by the news that arthritis patients have found mobility by movement and that arthritis has become less in the joints if they move that particular area. Personally I can't believe *'all'*. In saying that the Cardinal/moveable signs are Capricorn ruler of knees and bones, this made the revolutionary idea of bones joints moving very interesting. But, the other Cardinals are Aries ruler of muscles, Cancer ruler of stomach & breast area, and Libra ruler of kidneys, back/lumbar region. If arthritis can be cured by movement we could maybe then get dialysis patients to have massage around their kidney area, again it mightn't work for everyone, it could be too painful, a cancer patient of the stomach could find massage also helpful and also around a person's nipples.

When I go by the Encyclopaedia of Medical Astrology it uses all aspects – Aries/1st house/Mars or Scorpio/8th house/Pluto and for example Aries rules pimples because they're red and red means the blood under the rulership of Aries is moving but being blocked; sudden eruptions which we have to consider Cardinal/Active moving we'd also have to consider the word 'sudden' which indicates Uranus, sometimes we can actually feel a pimple developing under the skin which wouldn't include Uranus, but it gives the sense of movement. I have found by being in consultation with a person I can get to the problems faster. But, in saying that I haven't always had that report as mentioned in N.S. Chart.

When making a formula and in deciding what product to use I confirm my thoughts with the Runes, the reason I can use these is because my Mercury/Sun/Neptune are conjunct and makes a positive aspect to Pluto; progressed Saturn is also now only a few degrees away from my Mercury, over the years with the way my mind process things I've found I can get a correct idea faster; even if I'm looking for something that helps arthritis I'd look to what symbol rules Capricorn, goat leads me to goat meat or goat oil.

Now to preparation – if we have Mars - a fire sign, in Scorpio a water sign, in the 9th house another fire sign. The signs here are telling us that the product will need to be a Mars product a Scorpio product and a 9th house product. The Triplicities are telling us the products need to be cooked; for Mars it could be sheep meat, with the fire signs it would need to be cooked in water (Scorpio element) but, Scorpio is telling us the Water Scorpion, which I've never found, so I look to other things the Scorpion rules i.e. pollution and souring a product or it rules the masses so I'd look to a food where they are in the masses and small or humongous i.e. ocean prawns. To me Scorpio is also the color of indigo so I'd consider something purple, the skin of eggplant or octopus ink, further Scorpio rules the Reproductive System and hormones associate with that are, so anything that is looked for is because this area is being affected. Following, we could also use an orange carrot and the rice, the orange or purple carrot needs to be dried with heat and so it could be fried or grilled, a little bit of history tells us that rice is dried in the Sun and so it has already undergone one of the processes. The purple carrot needs to be prepared by boiling which combines the effect of

Mars and leave it to be soured, a property of Scorpio, and an acid. Finally, we combine them after leaving the purple carrot to sour and straining. This is just an example of how we/you can find and make a formula. The products are real for the problems I've suggested and it should not be taken lightly.

Remember though in a chart the combination of food and process only applies when planets are aspected, negatively. When there is a conjunct I'd use all the planets products if making a negative aspect to something else.

Chapter 12 - Explanation to Formulas & Examples of Formulas Created

MARK S. - PERSON IN A COMA

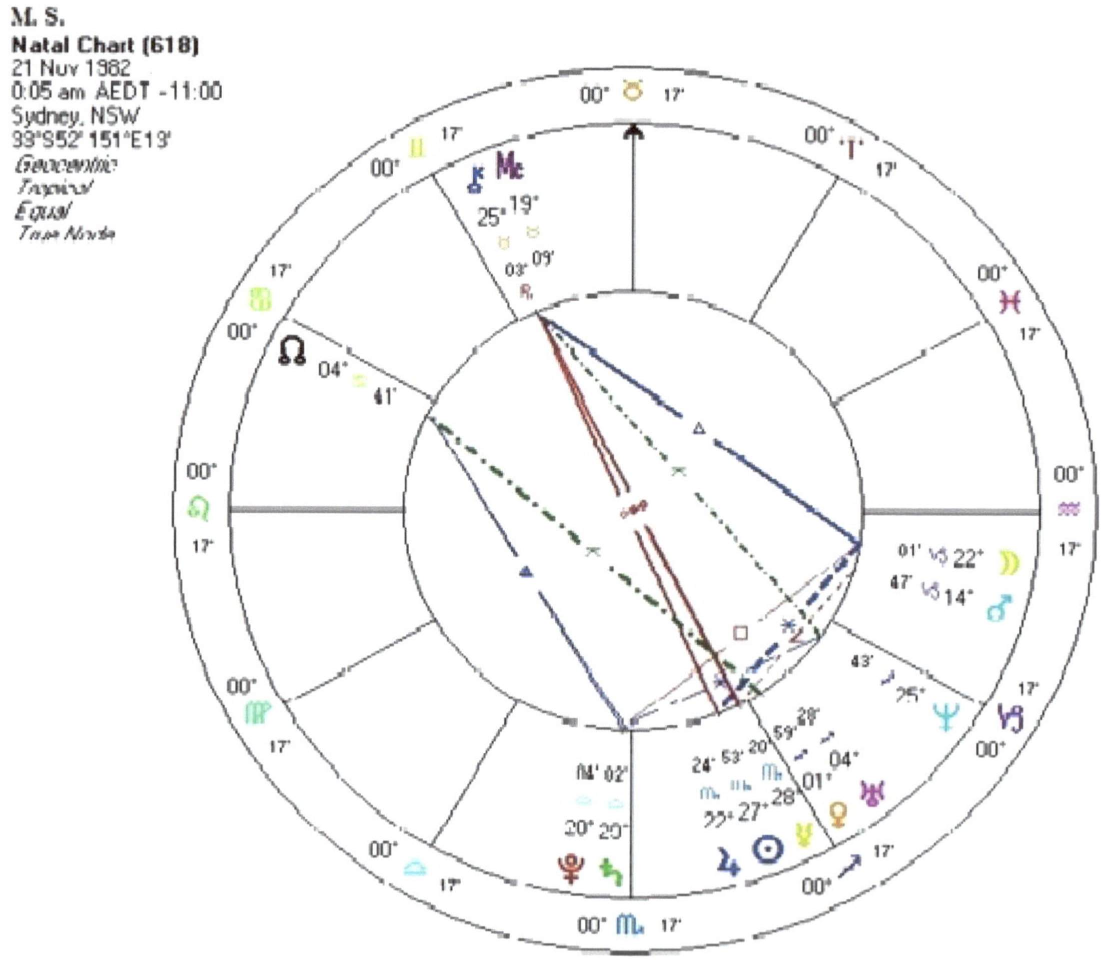

CHART 8

Above is the Natal Chart of Mark when he was 29 he had an accident whilst in a televised competition, which left him in a coma, to this day I believe he's still in that coma.

I'm showing the calculations in the chart that led me to several formulas; it would take a whole lot of guts to trial it. But, I've tried so many of my own formulas that work on a usually prolonged temporary basis, as planets change, that I was confident in making it up. Whilst I had reservations in giving his wife the process in which I found the formula, knowing it would only be useful for him as it is based on his personal data for time, date & place for the Natal, Transit & Progressed Charts. I felt I had to give her something as she had paid me $600 for the consultation. If you look at my chart you will see I have a lot of Pisces influences either by Pisces itself or Neptune, but not so much the 12th house, and there are quite a lot of blockages; and this is what causes me problems in doing quantities of analysis'. With Mark It was an innate feeling I had, that I had to get started on formulas. I am at no way confident when having to put to a guardian something has to be digested especially when it smells like something regurgitated, whilst I'd do it to myself if required. As you can imagine the case was quite challenging, and the formulas I've suggested are individual to the client and I had problems giving his wife what I'm about to give you as you can imagine it's my intellectual property, and at some time could be used for someone else, I hope that anyone using my procedures will accord me all rights warranted as having the intellectual property of the forethought.

For myself I had to have the proof that reasoning was correct and at the interview the wife and mother wanted me to see him whilst I was determined not to. In our consultation I brought to them the areas I'd found prominent in his chart. The thing was if I'd seen him they could have said my analysis was based on what I saw. I've always said that astrology was a science based on the planets and stars where you only need an accurate date of birth to begin; with preferably the correct time of birth all else follows, below is what I explained to them. The formulas were never given to him because one in particular smelt extremely bad, it took the longest time to prepare because it had to be fermented. I also suggested that they talk to the head doctor about what I'd found and about the formulas I felt would be required. Below is a detailed outline on what I saw and why I created the several formulas I made. You will see that I've been selective on what aspect I've used and only required the family to act in keeping Mark cool especially around the Transiting period outlined in bold. When looking at the digestion of the fermented crab for his stomach, even though it would have been feed through a tube into his stomach, afterwards I wondered if I could have made into an oil rub.

M. S.
D.O.B. 21 Nov 1982 T.O.B. 0:05 am
Place – Sydney Australia
Progressed date - 27 Nov 2012

The following is giving explanations to how I saw the body malfunctioning with referring to medical astrology, and the planets, constellations and house basics along with their aspect.

Explanation

From M.S's charts I explained what I saw and that was the hormones were being heated causing the groin to be hot, and one of the problems was the hydrochloric acids of the stomach where blocked by the circulatory system, the circulatory system are the fluids carried through the body it involves the electrical system but is prevalent in the intestines. It also showed that groups were upsetting his mind, and that there would be some sign of this physically.

Astrological Explanation

1. Natal Sun 27° Scorpio Square Progressed Moon 24° Aquarius 6th house
The Sun in Scorpio means the hormones are being heated and causing the area to be hot in the groin, what is squaring this area by progression is the Moon, which is moist has to do with the Hydrochloric Acids of the stomach being blocked in the Circulatory System ruled by Aquarius. The Moon is ill placed in Aquarius in the 6th house (a talkative house) meaning family in groups could be physically and mentally, upsetting him, therefore, preferably only 1 or 2 persons at a time. The Moon also rules wives, mother's family and the emotions, physically the stomach and mammary glands. Aquarius rules Groups, the Circulatory system and the Electrical nerves, therefore -

Suggested action
Restricting visitors to 1 or 2.

Explanation

Mark's liver was acting unpredictably in purifying his blood this would cause his skin to appear blue and yellowish and there could be breathing problems because of the focus being on the stomach. It also appears that the hydrochloric acids are having a negative effect on the Liver.

Astrological Explanation

2. Natal North Node 4° 04 in Cancer Quincunx Uranus 06° 17 in Sagittarius
This means that the Liver ruled by Sagittarius is acting unpredictably because of the planet Uranus, Sagittarius also rules the skin/liver would cause the skin to appear blue and yellowish. The stomach could appear white with North Node in Cancer. Sagittarius works against the action of the brain for it naturally opposes Gemini, which is an air sign requiring oxygen. The Cancer sign being 4° prior to the progressed Uranus shows that the hydrochloric acid is playing negatively on the liver. North Node acts much the same way as Sagittarius it expands everything and makes it noticed and so I put a Silica bases to it. The Cancer sign normally makes a 150° aspect to Sagittarius these influence is affected when there are planets in them, any 150° aspects are usually underlying problems.

Suggested formula
Calcium Ferr is what I see in Shell of Crabs. Primarily the Crab is the symbol of Cancerians, I suggest it to be rubbed on ankles which is ruled by Uranus, the Blue Crab which has a blue claw – I've found cooked at Shellharbour Fisheries, can be grinded together keep refrigerated. The blue is the colour for Uranus and i imagine the claw as the ankles of the crab having a relationship to the colour

Explanation
There could be too much attention paid to the stomach whilst the muscles in the circulatory system which bring fluids to the heart and around the heart are becoming fixed. Potassium would be going into the ankles causing them to heat up and possibly causing cramping. The circulatory system requires salt and in

fact contrary to popular belief salt helps the stomach produce hydrochloric acid, how it's administered is found in suggestions.

Astrological Explanation

3. Natal North Node in Cancer 4 ° deg. 04 12th house quincunx Progressed Mars 8 ° deg. Aquarius in 5th house

North Node in Cancer means attention is drawn to the stomach but what mightn't be obvious is the muscles Mars influence in the circulation is causing problems to the heart 5th house, this would be with salts going into the stomach. Also it could mean too much water in the ankles and a warmth caused through Potassium being focused in the ankles heating up the Circulatory System.

Suggested formula

Acupuncture around the ankles could help. Mars rules needles. And, as the patient can't talk, where there is excessive heat placing an acupuncture needle in that area would help the potassium's move. If a family member is doing it they'd need to feel for indentations and only 5mm and should be kept away from bone areas. This can happen anytime family members are there. This could also be performed around the ankle as it's ruled by Aquarius.

Explanation

There is an obvious obstruction where the sugars of the body are causing the stomach not to act effectively, causing calcification around the knees and in the bones

Astrological Explanation

4. Progressed North Node 4 deg 15 Cancer 10th house Opposite Progressed Venus 09 deg 08' Capricorn 4th house

Cancer sign rules the stomach it's an active sign when there is a planet opposing the North Node it shows that planet to be acting untimely being affected by the South Node and in this case it is Venus in Capricorn – Venus implies sugar a carbohydrate is causing the stomach not to act properly. Sugar is ruled by Venus and is likely to cause calcification when the planet is in Capricorn. An interesting comment was made by the mother regarding his legs cramping is in the final notes.

Explanation

The brain requires massage; the brain not being massaged would mean there could be air in the bones that are not being released.

Astrological Explanation

5. Progressed Mercury 14 ° Capricorn

Mercury rules the brain; Capricorn is an active sign and would mean working in a physical sense. It also could mean that there is too much air in the bones, and possibly the Marrow of the bones. Mercury being an air sign. Capricorn ruling the ligaments/bone/knees and marrow inside the bone it is also a cardinal/active sign which differs from Mercury which is mutable/talkative and air sign.

Suggested action

Massaging around the head should be beneficial, as the head is covered by bone and is continually rejuvenating itself massaging should help to promote the air through the bone structure. The other area that could benefit with massaging are the knees. An alternative could be putting air on the knees for brief periods

Explanation

There can be seen infection, if red and heat can be felt in the bones where acupuncture could help, it's caused through heat of the blood affecting the calcium of the bones and minerals and the marrow unable to reproduce, for suggested treatment refers to suggested formula.

Astrological Explanation

6. Natal Mars in Capricorn

Mars is in its detriment in Capricorn it can mean being aggressive in the field of work, where Capricorn is already an active sign and not one to sit on its laurels. This involves heat in the bone structure. It can mean that the Marrow in the Bone is heating and therefore the natural regeneration of the bone system is having problems rejuvenating itself, Mars also rules the Blood and it's action of using the muscles gets the blood flowing, this will be another area needing to be looked at, but the treatment below intends to help this problem.

Suggested formula

a.

Recommended formula - Time: 9 – 10 Tuesday Best day to work on muscles.

¼ tspn of Crushed Wheat an earth/air mix with 1 tspn of Hot Chilli Sauce this needs to be rubbed into the back of the head – Cerebrum and Cerebellum

b. This aspect shows acupuncture on the head at the Chinese meridian Du 20 useful. I'm suggesting only one needle, which a family member could learn to do, laying the needle flat facing horizontally between each ear. (It's the last area to mend in the head as a child) where there is an indentation in the top of the head, I have provided diagrams to show how to find meridian points. Ex 6 meridians actually circles the area in the top of the head, Ex 6 point can be used after you see a reaction. You could also place heated wheat around the knees.

Astrological Explanation

7. Planets in the House position affecting the Brain – Best day to work on Kidneys is Friday

3rd House aspects Natal Pluto 28° deg. 04 Libra

Pluto in Libra means an excess amount of sugar in the Kidneys. This would mean the Kidneys would be finding it hard to get rid of the Sulphates It could be also depleting the air to the kidneys. Bringing this to the doctor's attention as patient could inhale a very minute amount of nitrogen, which is a substance of Pluto's

Transiting Jupiter in Gemini 21-1- 13

For this day it shows the influence of Jupiter possibly expanding the air in the lungs. These are areas that need to be monitored by a doctor.

Suggested formula

a Inhaling Nitrogen .5 mm alternately 3 drops of Ammonia in 2 tbspn of Saline, 2 tbspns of water in Asthma machine, probably be all that's needed.

Astrological Explanation

Transiting Mars 20° 03 in Aquarius squaring Jupiter 22° 24 in Scorpio

This aspect for a few days around 21-1-13 will need to be kept an eye on.

this can mean that the Air to the Brain is heated and could at times be blocked by being dried out with too much heat, and constant monitoring should take place.

Suggested action

a. A damp cloth on the Brain and temporal area recommended.

Explanation

There could be seen calcification or excessive calcium in the kidneys.

Astrological Explanation

Saturn 29° Libra 3rd house.

d) Saturn is a planet which is naturally working when in Libra it can be found that it works against the action of the kidney, doctors will probably find the kidney working too hard.

Suggested action

a. Saturn in Libra shows a Potash or crushed Calcium Phosphate they are ruled by Saturn, and would probably be useful on the Kidneys as this planet would be probably restricting the Air both in the kidneys and lungs and required action to the Brain. Using the above could breakdown the Calcification that has been built up. A form of Air Brushing would be best for the kidney area and massaging it into the Lung area.

Explanation

The blood from the liver could be forced into the groin instead of kidneys and bone marrow.

Astrological Explanation

8. Natal and Progressed (N/P) Jupiter in Scorpio in 4th house

Jupiter rules the liver whilst Scorpio rules the reproductive organs, the 4th house is normally that of Cancer which rules the stomach this shows the purified blood has been forced into the groin instead of the kidneys and bone marrow, which is used to rejuvenate the bone. Resulting with problems with the stomach.

MOTHER AND WIFE'S INFORMATION

In the consultation the areas I focused on caused them to inform me that he was being intravenously fed through his stomach and so there was a lot of focus on his stomach which I had no knowledge of and very surprised about. I stated he was getting upset with crowds and his mother said he appeared agitated at visiting times. I said in his own way he was talking and his mother said 'yes, he made murmurs'. When I said there were problems with his knees they informed me that he'd recently had surgery on his hamstring because his knees were cramping up.

The following is another chart combining the transits at the time I spoke with Mark's mother and wife, with the progressions at the time of the interview, with the natal chart.

M. S., 21 Nov 1982 - Progressed 27 Nov. 2012 Flow Chart

	Deg	Const	Hse	Quad	Tripl	Area	Area	Negative Aspects	Conj	
Sun	28° 20	Sag	4	Fixed	Fire	Heart	Liver		Nept in Sag	Lymp/feet
							Hips		/Sun in Sag	Liver
Jupiter	28° 52	Scor	3	Mut	Fire			NN - Can		
Pluto	28° 59	Libra	2	Fixed	Water			210	Pl in Lib /	
Saturn	02° 01	Libra	2	Card	Earth			Mars in	Sat in Lib	
								Aq		NN Liver
										Cancer 10th
NN	04° 15	Cancer	10			Liver	Stomach	NN - Can	Ven in	dried Milk
SN		Cap	4					150	Cap/Mercury	Mars in Aq 6th
Uranus	06° 17	Sag	3	Fixed	Air			Ur - Sag	in Cap	Wheat air grown
Mars	08° 02	Aqua	5	Card	Fire					
Venus	09° 42	Cap	4	Fix/card	Ea/air			NN in		
Mercury	14° 51	Cap	5	Mut/mut	Air/Ea			Cancer		NN Liver minced
Moon	24° 19	Aqua	6	Card	Water			opp		in water
Neptune	26° 49	Sag	4	Mut	Water			Ven in		Cancer 10th dried
								Cap		Milk
										Ur - Iodized Salt
								Saturn		Sag in 3rd Liver
								in Libra		air dried
								sq Ven		
								in Cap		Sat - Flour
										Lib -
										Strawberries
										Ven
										Cap - Oil

When looking at ingredients the foods required are only those where there has been conflict within the planets/houses or constellations, whilst every sign has a food, Sagittarius might be the hardest, but many forms of nuts have silica in them but, I've always found Liver the anatomy part it rules the best.

In the following pages I've expanded on the above as each constellation/house have a triplicity/quadruplicity and are either alkaline or acidic, it's an extension of the above chart I gave, but using the flow chart of the above based on the degrees and minutes, you will find that some planets are seconds apart, and is a consequence of having the correct time. Below helps with colours when cooking, refer to Color Chart for Constellations, House & Planet Color, obviously equally important is whether something should be alkaline or acid, if a product isn't naturally acid or alkaline it may need to be fermented or made into an alkaline and sweetened down, but remember whilst I have every planet/house and constellation there we are only looking for the health aspects that are conflicting with the charts

Constellations	Quad	Trip	Alk	Acid	Hse	Quad	Trip	Planet	Quad	Trip
Sagittarius	Mut	Fire	Alk		4	Card	Water	Sun	Fixed	Fire
Scorpio	Fixed	Water		Acid	3	Mut	Air	Jupiter	Mut	Fire
Libra	Card	Air	Alk		2	Fixed	Earth	Pluto	Fixed	Water
Libra	Card	Air	Alk		2	Fixed	Earth	Saturn	Card	Air
Cancer	Card	Water		Acid	10	Fixed	Card	Nth Node		
Sagittarius	Mut	Fire	Alk		3	Mut	Air	Sth Node		
Aquarius	Fixed	Air	Alk		5	Fixed	Fire	Uranus	Fixed	Air
Capricorn	Card	Earth		Acid	4	Card	Water	Mars	Card	Air
Capricorn	Card	Earth		Acid	5	Fixed	Fire	Venus	Fix/Card	Air/Ea
Aquarius	Fixed	Air	Alk		6	Mut	Earth	Mercury	Mut/Mut	Air/Ea
Sagittarius	Mut	Fire	Alk		4	Card	Water	Moon	Mut	Water
Sagittarius	Mut	Fire	Alk		4	Card	Water	Neptune	Mut	Fixed

M.S. Natal /Progressed & Transit Flow Chart

Natal/Progress & Transit	Degree & Minute	Constellation	Planet	House	Aspect
N	27 53	Scorpio	Sun	4	30° P Jup/Nept to Sun
N	28 04	Libra	Pluto	3	⚹ N Pluto / P Sun
N	28 20	Scorpio	Mercury	4	
P	28 20	Sagittarius	Sun	4	
P	28 51	Scorpio	Jupiter	3	□ P Jupiter - T Sun
P	28 59	Libra	Pluto	2	⚻ P Pluto/N Saturn -
*					T Neptune
N	29 04	Libra	Saturn	3	
T	00 36	Aquarius	Sun		
T	01 40	Pisces	Neptune		□ T Nept - N Venus
N	01 59	Sagittarius	Venus	5	
T	02 00	Aquarius	Mercury		
P	02 01	Libra	Saturn	2	□ P Saturn - P NN
P	04 15	Cancer	Nth Node	10	
T	06 30	Gemini	Jupiter		☍ Jupiter – Uranus
P	06 17	Sagittarius	Uranus	3	
P	08 02	Aquarius	Mars	5	
P	09 42	Capricorn	Venus	4	
T	10 00	Capricorn	Pluto		
T	10 47	Scorpio	Saturn		
T	14 12	Capricorn	Venus		
P	14 51	Capricorn	Mercury	5	
N	14 47	Capricorn	Mars	6	
T	17 00	Taurus	Moon		☍ T Moon – N Jup
					□ T Moon - T Mars
T	20 03	Aquarius	Mars		□ T Mars - Jupiter
					□ T Mars - T NN
N	22 01	Capricorn	Moon	6	
N	22 24	Scorpio	Jupiter	4	□ N Jupiter - P Moon
T	23 47	Scorpio	Nth Node		□ T NN - P Moon
P	24 19	Aquarius	Moon	6	
N	25 23	Sagittarius	Neptune	5	
p	26 49	Sagittarius	Neptune	4	

NEVILLE R. – STROKE PATIENT ANALYSIS & TREATMENT

CHART 9

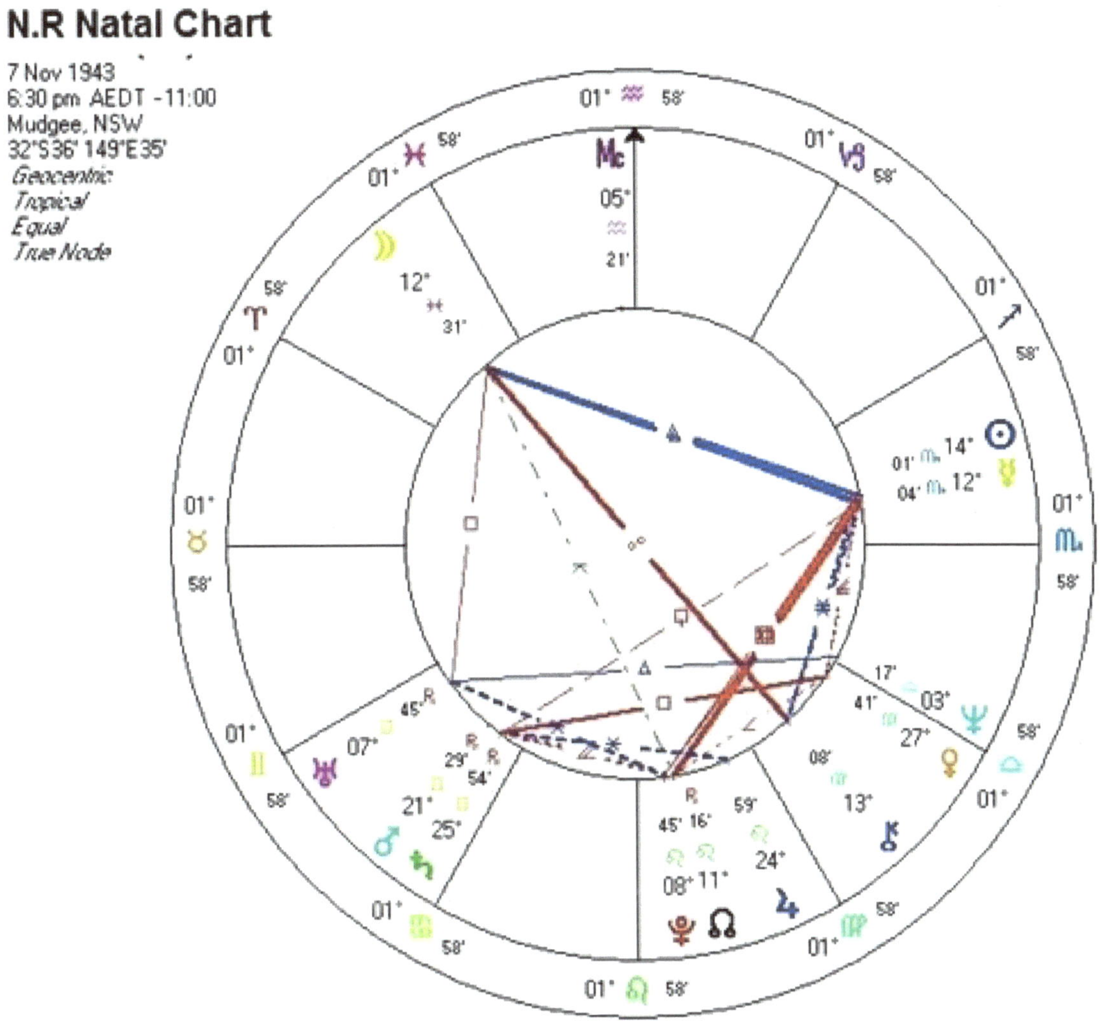

The above Astrology Chart is the Natal Chart of a person who had a stroke either playing or after playing football, his age was around 27, at the time of volunteering for astrology formulas he was 59. Over a 3-day period with following the Natal, Progress and Transit great progress was made where the therapy he received was gentle but proved strong, at the end of three days he was almost walking with both feet on the ground, his posture had improved and every month he improved more, it lasted 15 months when a lump on the other ankle appeared, unfortunately I wasn't consulted, but I did have a feeling that a negative reaction would come about when I saw him embarrassed and abused at his favourite club, after all he'd been paralysed the left side of his body for 32 years.

When we first started the therapy I needed to explain to him that I didn't know how his body would react because in Astrology in medicine there are four major concerns which relate to signs that are opposite and squaring each other concerning an illness, this has been supported over and over again by the medical profession, and, because I didn't have a computer mind I wouldn't know what would happen in a months or a years' time, but, I stressed if he had any problem he had to tell me, and so I explained the opposite and squaring signs Neville R. would need to be aware of. As a Scorpion his sign ruled the reproductive areas, opposite is Taurus which rules the Thyroid/throat and neck, squaring is Leo which rules the heart, and Aquarius which rules the circulation and ankles. Maybe I just didn't emphasis enough. As he was a Scorpion I was scared as he had 8 areas in fixed constellations and houses. As they were in fixed areas I felt the organ may have shrivelled up with the enzymes not getting to them. I explained that I believed that the strain of an organ suddenly working could affect any of these opposing and squaring areas, and that I wanted him to tell the doctor what we had done, and, what changes he found in himself.

He admitted years later he thought it was a game.

Unfortunately, when I again saw him I'd put myself into a bad financial dealing and I was over stressed and not handling things very well. Where my word came over all common sense and self-preservation, in fact nothing was making sense, I had still been trying to get formulas reviewed by Ministers and just as Neville R. had told me he thought working with astrology was stupid I believe that's what I thought Ministers were thinking. Later I explained to him that the 3 days we spent over the Easter Weekend I'd put my heart and soul into it. I was working on hours in his chart, when to do the acupuncture, the massage, and I was making up formula's on the spot. I was buying the foods that I had no money for, because I wanted to help him. I'd also explained to him that I had taken many a formula to a Minister and asked for reviews, where they would be forwarded to the government research department, to only find two years later they were patented. Throughout my life I was plagued with computer problems and when I'd addressed the Minister for copies of my letters they were appropriately lost.

What I couldn't understand is why Doctors didn't know how to treat a lump on a limb, which could only be fluid and so he reverted to limping. He said 'he couldn't afford to have his good leg fail him'.

The sudden abuse he had at the local bowls club, would be seen in his chart by possibly a Mercury /Uranus aspect or an associated form. There can already be seen quincunxes with Uranus in Gemini 7° 2nd house and Mercury in Scorpio 12° 7th house. To be convinced a progressed & transit chart would need to be added to his natal, where I've done an estimated progress below.

The situation was he was showing me how to play bowls when he was abused, which leads me to Uranus. unpredictable and Mercury rules the neighbourhood. In checking the transits whilst it can't be exact 15 months later Mercury was conjunct Uranus and as the progress Sun moves 1 day a year it would be making health problem by a quincunx aspect. As such this shows how treatment can be altered and how astrologers can look to future aspects, for altering formulas.

Now to the formulas, as always the formulas are made up to find the food types, suggestions can be seen under Foods in this book. For Neville R. a few were pre-prepared and fungi was made when I saw Pisces had planets opposing, surprisingly I'd made these for myself and was amazed I could use them. The therapy itself was on the spare of the moment, and other treatment included alcohol/massage/ hypnotherapy/acupuncture and colour, all of these were based on aspects. It was the first major therapy I had done which required him to keep close contact with his doctor. To see him walk straight brought unexpected jubilation.

Below is the important part of analysing what is required for a formula.

To begin this, I do the *flow chart*. As with Mark it's in order of degrees, you can begin anywhere but I begin with those closest to the Sun. Whenever deciding on colour I've tried to combine the colour of not only the planet but house and Constellation and the aspects of close planets. Please see 'Colours' for more explanation. This chart helps to see immediately what planets are close to each other. In charts I look to conjuncts with 3°- 6° orbs as with aspects 0°, 90º, 180º, and 150º angles. As with Mark I've done the 3 calculations with Natal, Progress and Transiting, remember that transiting planets are fast moving planets and may only last a short time. The further out planets of course last a lot longer and I usually don't double up because e.g. Pluto may have only moved if 1 degree. The transits must be paid attention to as the formula must be altered once the planets moved on and that food type taken out.

In the flow chart I've repeated the Triplicities and Quadruplicities, these are important in deciding on a colour and way of cooking or preparing the formula. A fire sign indicates orange/yellow and red, and on an obvious level - heat. This is easily seen with a lighter, the top flame is yellow the deeper flames are orange and red, there is always a deeper analogy because when I lit the flame I saw the base of the flame was blue (blues the opposite colour to yellow under Colours, and something has to cause a spark which is the gas and the friction) without air, nothing would be set alight, you wouldn't have a flame at least in the lighter but the flint is another additive. In astrology yellow and blue are fixed colours, they are also known as primary colours along with red and green again refer to Colours. Fixed elements are part of the Quadruplicities and this makes me to believe in the primary colours there are four, not 3 as believed.

In health the fire signs cause a temperature. If there is a blockage in a fire sign the homeostasis of the body would become unbalanced, homeostasis is the regulation of the temperature & enzymes; it means the body wouldn't be able to keep warm or possibly the opposite extreme heat this depends on degrees and whether the planet's degrees are before or after.

Remember he was fine till the aspects 27 years later the flow chart is based on his Natal because this doesn't change, the Progress chart is based on the day month and year and hour we began his treatment. The only foods I'd require for a formula are those making a quincunx, square □, or opposition. There is also the problem of planets being in detrimental signs which they are normally opposite/square ore quincunx which I've also outlined. Below shows the foods I used and how I came to combine and create them.

Neville R. Flow Chart for D.O.B.

Planets	Deg	Trip	Quad	Acid/	Hse	Trip	Quad	Acid/Alkal.	Const	Trip	Quad	Acid/ Alkal.	Aspects	Color	Food
Mercury	12°04	Air	Mut	Alk	7th	Air	Card	Al/Ac	Scor	Water	Fix	Acid	Ur ⊼ Merc Pluto/Leo □Merc/ Scor	Purple	Celery/Air, Polluted cooked Drink water
Moon	12° 31	Water	Card	Acid	11th	Air	Fixed	Alk	Pisces	Water	Mut	Acid		White/Blue	
Sun	14° 01	Fire	Fixed	Alk	7th	Air	Card	Al/Ac	Scor	Water	Fix	Acid	Sn/Mars	Yellow/Red, Yellow/Black	Tomatos Cooked with seed + Hot Sauce
Mars	21° 29R	Fire	Card	Alk	2nd	E/Air	Fixed	Acid	Gemi	Air	Mut	Alk	Mar incon merc	Red/lgt Brn/Silv	Massage head above ears
Jupiter	24° 59	Fire	Mut	Alk	4th	Water	Card	Acid	Leo	Fire	Fix	Alk	Vir Gem	Yellow/whit/Yell	
Saturn	25° 54R	Earth	Card	Acid	2nd	E/Air	Fixed	Acid	Gemi	Air	Mut	Alk	Saturns ruler ⊼ Gemini		Massage Celery Oil on Arms
Venus	27° 41	E/Air	F/Car	Alk/ac	5th	Fire	Fixed	Alk	Virgo	E/Air	M/M	Acid	Saturn □ Venus	Brown	
Neptune	03° 17	Water	Mut	Acid	5th	Fire	Fixed	Alk	Libra	Air	Card	Alk		Yellow/Pink	
MC	05° 21				10th				Aqua	Air	Fixed	Acid			
Uranus	07° 45R	Air	Fixed	Alk	2nd	E/Air	Fixed	Acid	Gemi	Air	Mut	Alk		P.Blue	Salt
Pluto	08° 45	Water	Fixed	Acid	4th	Water	Card	Acid	Leo	Fire	Fix	Alk	Pluto ruler □ Leo	Purple/Yellow	Potato with Oil for head

The following chart shows the stroke patient Neville R.'s Natal Chart in the inner wheel and Progressed chart in the outer for the 1970 when the stroke took place. I don't believe it requires a *flow chart* only that it would help to show in another form the planets degrees and their elements. It would be a formality to look at this if you wanted to know what could have affected him at the time the stroke took place, but it wouldn't be necessary for a formula.

CHART 10

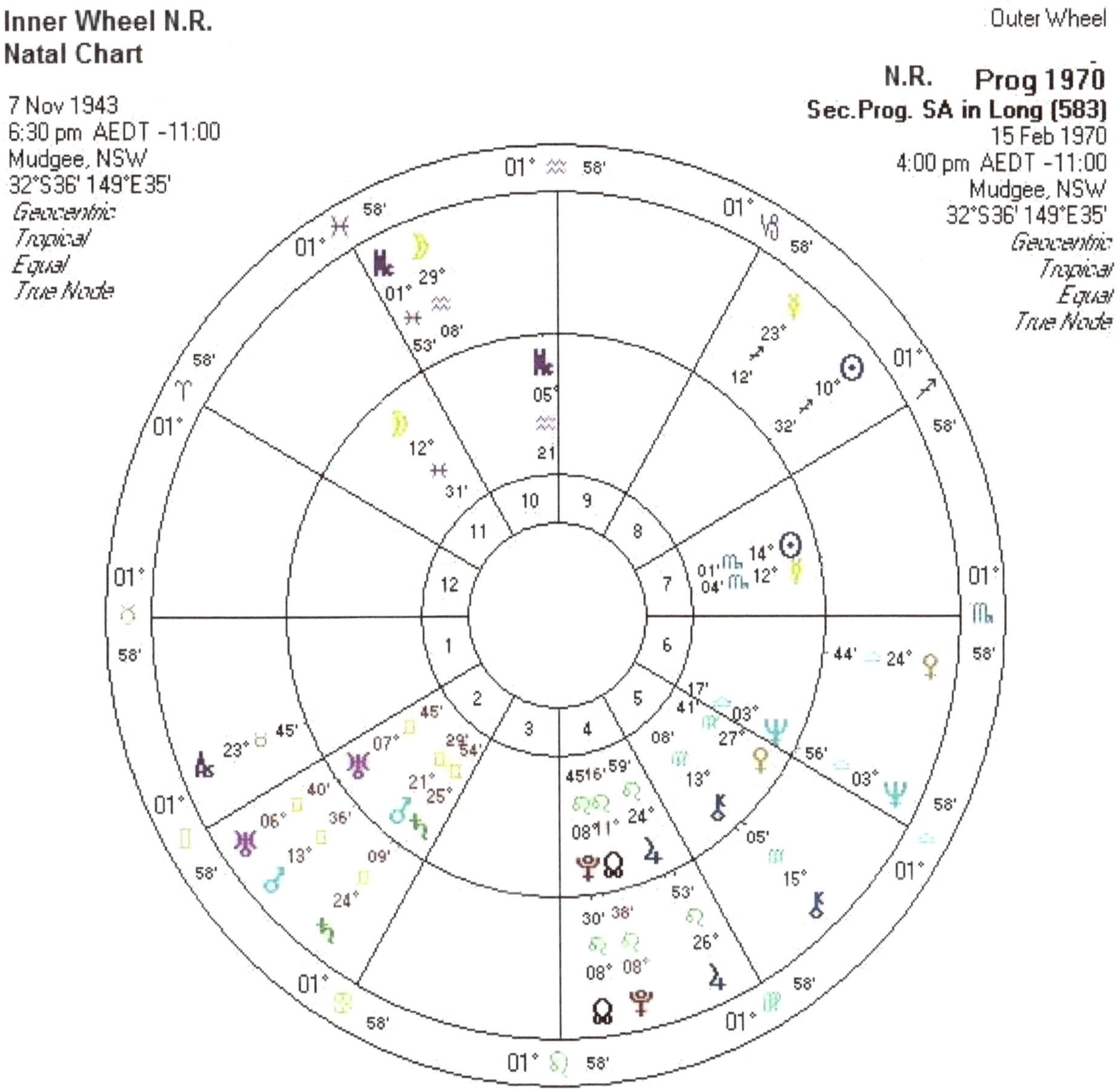

When we look at a person who has had an illness and in this case a stroke we have to consider what has happened and what has been affected; Neville R. couldn't physically move his left arm or leg, in fact one would expect that his organs to some extent had been paralysed on the left side as well, but then again because the brain is ruled by Mercury and the legs are ruled by Jupiter maybe they are the two aspects we should be looking at. I did expect problems with the eye and mouth, but, he didn't appear to have the later problem. The natal and progressed charts are accurate for the time. We can see natal Mercury in the fixed sign of Scorpio but in the active house of Libra working in partnership, we could look at the slower planets that were in Gemini the constellation ruled by Mercury and these are in the 2nd house fixed constellation usually ruled by Taurus.

Mercury still rules the arms which we can see in the Natal Chart as being 12º 04' Scorpio, Mercury in Scorpio already shows 150º if looked at in its natural sign of Gemini and so when met with other aspects is in a delicate position. The arms are an outer expression of the lungs in astrology and, if we look at Gemini the progress Mars has moved to 150º aspect to the Natal Sun and Mercury. Mars rules the red blood, head and muscles of the body. I would be concerned about exerting oneself in a physical manner, the Mars influence is hot and fiery, it is an active planet, but when in Gemini it's asking for it to be more talkative, an outcome of this would mean talking too fast, possibly burning up the oxygen. Natal Mercury is a degree earlier than the Mars degree so this is more likely to mean the gush of air would affect the muscles in the lungs, Mars ruler of muscles. Another quincunx 150º is Pluto to the Natal Moon. Pluto rules the groin and the hormones and its action is to be massive as the oceans. It usually has a good affiliation to the Moon but when making this angle we could be concerned about too much moisture or too many polluted fluids flowing into the lymph nodes. Pluto squaring natal Mercury and its own sign Scorpio would also bring about blockage to the Brain and Heart. According to the degrees it would first cause a problem with the brain ruled by Mercury which has a flow to the ☽ Moon being 19' later, which causes an already problematic area as earlier mentioned to Pluto. This is where the fluids could be too much. The Moon and Pisces are both water signs; Mercury in Scorpio is in a water sign. The Sun is in Scorpio and Pluto is squaring both Sun & Mercury. The Sun is the ruler of the heart and it shows that Mars was over-exerting the heart. With Mars 13º 36' and Sun 14º 01'.

The beauty of the North Node is that it is where the obvious focus is on. But, if they decide to look at an astrology chart and 150º aspects it is clear from a *flow chart* that we can look at a number of other areas.

The above chart shows how simply the problems of the heart can be corrected if they can take the emphasis away from death NN conjunct Pluto, Pluto is also remake or regeneration. From what I understand before the stroke he was a great sportsman.

The 29th March, 2004 was a milestone in discovery, it was the actual day we begun Neville R's. volunteered trials using the formulas made through what I could see in his chart, what was of great help was that he wanted to work on himself but more deeply his heart, Sun in Capricorn, 7th house meant in partnership. Mercury 21º 28'R in Capricorn in 8th House meant there would be deep research this was quincunx Saturn 21º 30'R Gemini on the 1st house cusp.

Whilst this was an area that needed to be looked at as it was a health aspect, 150º, individually it meant he was acceptable to his mind being changed. For 32 years he had never had any physiotherapy, and he admitted it to me that he just refused it at the time he left the hospital. They had told him he'd never walk and he did by his own recognizance, whilst struggling every step.

Neville R. had always enjoyed dancing, I had thought it was because of the stroke that when he did dance it was unpredictable and uncoordinated and, worrying. In looking at his chart when looking at the area that focuses on his legs we see Venus in Sagittarius, Sagittarius rules the Legs, this aspect is opposite to Mars one of the rulers of muscle conjuncting with 1º difference with Uranus one of the rulers of the circulation of the body it also rules the ankles, is the ruler of volcanos and unpredictable behaviour. When dancing it appeared that he wasn't conscious of fast and unpredictable movements. And, this would be natural with Mars and Uranus in the 12th House, which is ruled by Pisces, and probably why he could adapt to be crippled. The 12th house is said to show our own undoing; it's questionable whether this could be the case for Pisces or Neptune but there again it could mean depending on what planets are in it.
Now we'll look at Saturn 1st house, Sextile Jupiter ♃, Saturn shows the area he'd be working on, being in Gemini in the 1st house which is a cardinal or active house naturally shows he'd want to walk. This is a progressed sign and Saturn normally take 2 ½ years to go through a constellation

These works and records have been compiled over a period of 15 years. I'm not blessed with Jupiter in any favourable aspect except for freedom, and money certainly doesn't come easily for me; in some sense I was the silent partner in a marriage regarding finances, whilst I scrapped and saved for everything even when running a local paper nothing was easy, those problems came from often losing everything I wrote for an article and hadn't saved, I felt such an idiot. And when it came to everything I'd saved for it was all thrown away when a temporary separation took a more permanent arrangement and through illness I couldn't return. My husband saw no worth in anything I had created my Jupiter is in 8th house which has the characteristics of Scorpio being in a house position shows the area in which the action happens the actual constellation Jupiter is in, is that of Aries, which is direct active and straightforward; Partners are ruled by the 7th house with the characteristics of Libra. The Moon at 7° in Pisces is said to be a psychic sponge but at such time Pluto in Sagittarius was stronger for me not to know what was going on. Jupiter in Scorpio has given me the faith I have which I've mentioned earlier and that is the Magi Astrologer Priest often forgotten but were the followers of the star to Jesus ultimately the founders of Christianity; the Magi is believed to termed from the word 'magic' which appears now quiet irrelevant, astrology shows that some people have a gift of prophesy or the ability of illusion by a very swift hand, why I've gone into these areas I don't know all I do know is the depth in what I sometimes go coupled with my unpredictability, upsets my family and that which upsets the home upsets me. I know how and why it's caused and it would be with many people in my generation who have Uranus the feeling of unpredictability and freedom in the constellation of Cancer the home and family. Thou freedom without structure means you can lay down your roots in many places or not at all. My aspects are harsh *ones that oppose the brain (Mercury), the self (the Sun) and the intuition/psychic (Neptune) given that then I have a sextile from Pluto death destruction helps me to understand my life, to understand how things happen you do the flow chart which I've explained a number of times, based on degrees and minutes of the planets putting them in order from the Sun. I've been ever grateful for a man who was a guest speaker at Sydney Astrological Research Society, SARS teaching me that, I'm just sorry I can't remember his name.

Why have I waffled on, I guess it's to explain the 15 years in the making. When I began my research into the environment and not human nature, I desperately wanted a conclusive answer to whether I was on the right path, whilst I believe there are many paths astrology can take us it is the degree in which part of physics you're at the lessor will only limit you to the heights you can go.

In 2006 early 2007 I sent most of my formulas to ANSTO they were based on carbons, to this date I haven't checked if any have been patented. But, in 2016 I found new materials being made from carbons and silica.

NEVILLE R. PROGRESSED CHART FOR 2004
CHART 11

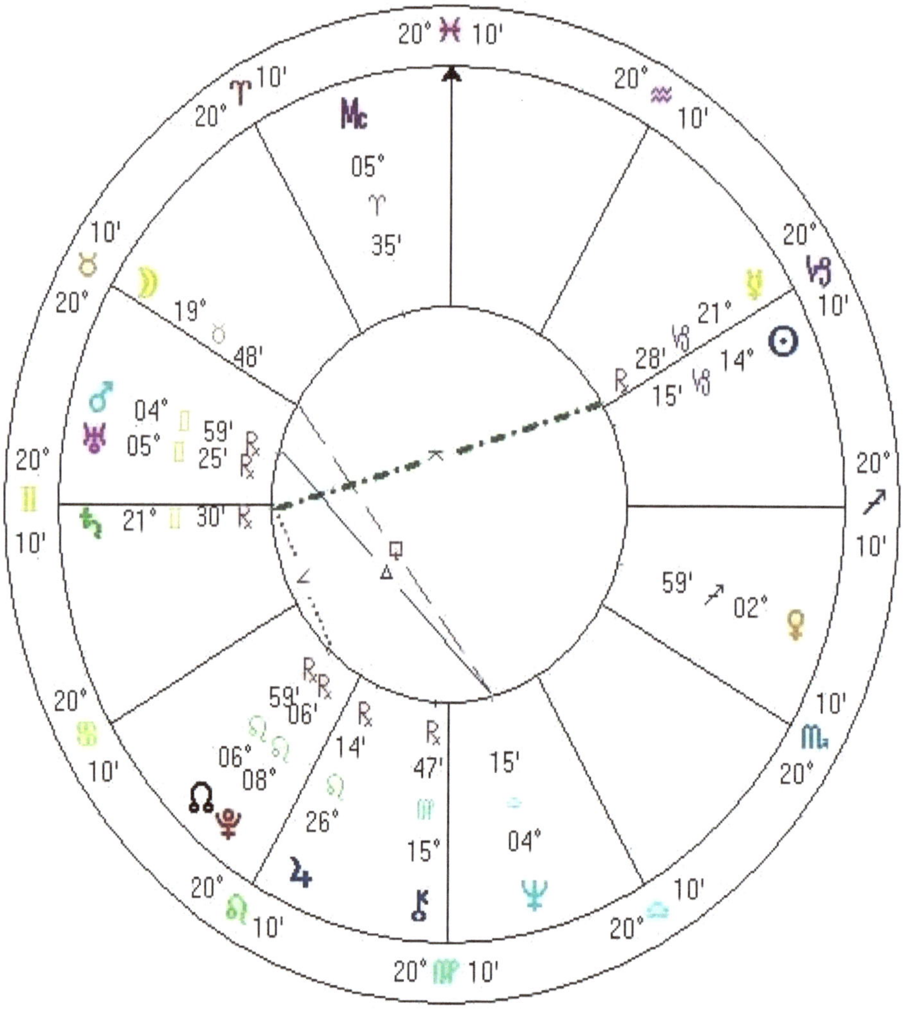

WAYNE R. – SKIN CANCER PATIENT NATAL & PROGRESSED CHART

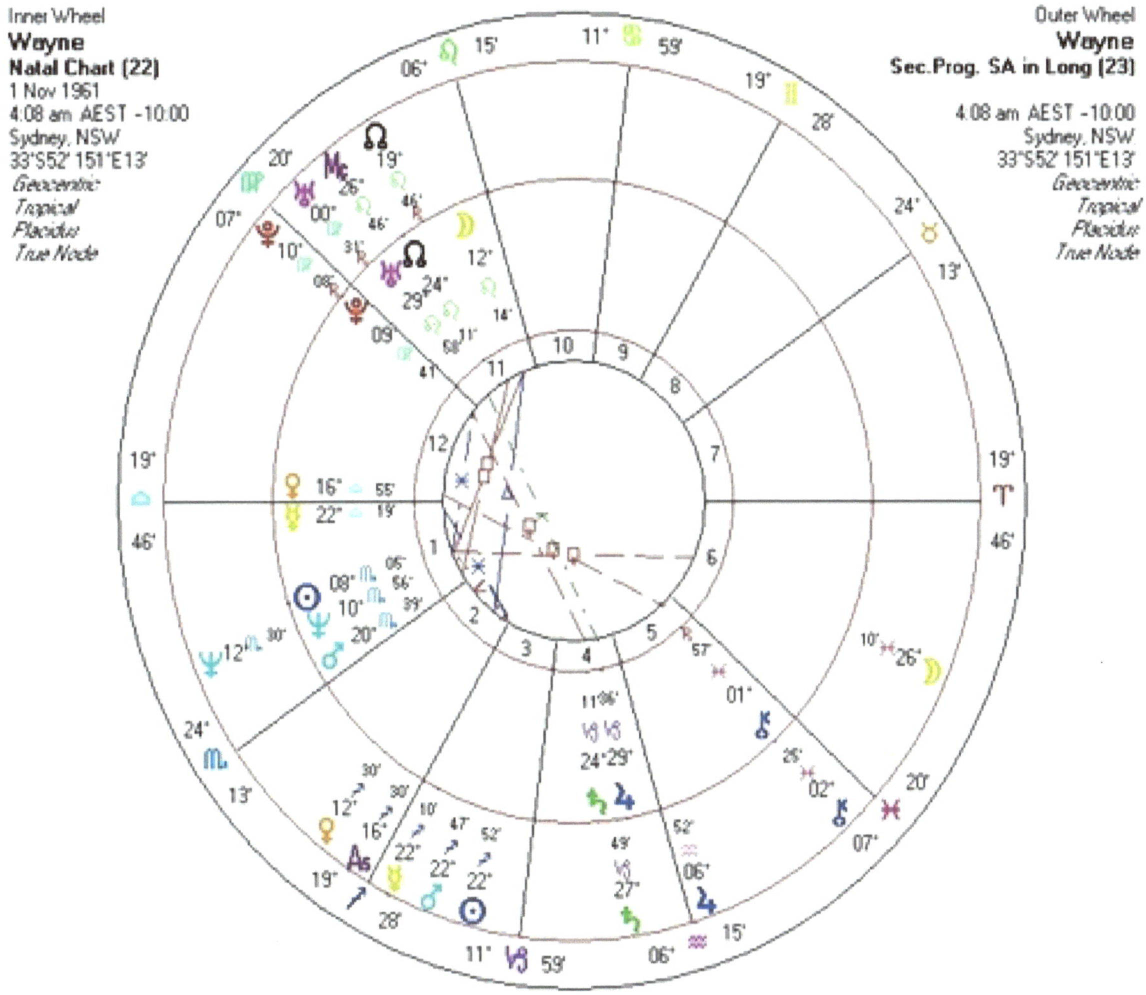

CHART 12

WAYNE R. – SKIN CANCER VOLUNTEER, ANALYSIS & TREATMENT

2007 was the same year I created a skin cancer formula for a gentleman I'll call by his first name Wayne, the heavens must have been with me. Of course when I first suggested I do a formula and there were his mates around they laughed. The problem is just as these men laughed at it people have lost the way of showing that by using what we have been told and have come true in the future it can go further. The initial question in my mind if God for another name gave us a predictive tool, a personality tool, why wouldn't he have given us a healing tool. My 2-year stint on 2BLU – Blue Mountains where I analysed current affairs issues and latest medical research showed to me God had. There was an acclaimed scientist who found mussels helped muscles and in Astrology by affiliation Aries/Mars/1st house rules muscles, to me it made sense that mussels could help muscles but it would have to be used in the correct way.

With the skin cancer formula, it was within days I had a formula, I was ecstatic, brain working/psychic working. I asked him to trial it and he said he would. The next time I saw him he had no sign of skin cancer I asked him to tell his doctor so that there would be some paper trail. I'd made a formula also up for his mate Dave S. I could take their sniggering because three of us were long term friends, Dave had suffered for a long time back pain and when he'd mentioned it before I thought it was temporary back pain, but when I took it seriously and made up a formula, he also came back and said it had worked.

These formulas may have only taken 56 hours to create but with my erratic emotional characteristic it was draining. And there again as what happened above happened again what I feel I've made from nothing was thrown away. For some reason they felt they couldn't defend keeping it, I presumed it wasn't their house. Strangely neither either of them ever had anything in their life, which was probably why they couldn't defend keeping them. It goes to show money isn't everything because at that time their support meant a great deal.

When looking at a chart you can also look at the situation surrounding people. Wayne was living with a family on a farm helping to raise wildlife and farm animals, but, I had felt and probably mentioned it to Wayne the philosophies of the family could possibly have caused the skin cancer, I say philosophies because Jupiter in astrology rules philosophies of others and in the anatomy the skin & liver. Where in health Jupiter was quincunx Uranus. But there's more which I'll go into when describing the formula's

In doing his mates chart who was nearly 40 and still lived with his parents, the astrological understanding to this is the back is said to have dual planets associated with it Saturn the bones, the Sun whilst it rules the heart it is closely associated with the spirit, but pride plays on structure. Figuratively it rules the Self, Son and the Father. I say the spirit because when I was told my mother had died. I sat with her and gradually followed her body going cold, and felt it was her spirit leaving her. It may have been the close Neptune association to the Sun, I'm not sure, in fact it's the first time I've mentioned it.

Getting back to the aspects, whilst there's a possibility the Sun has sort of a spiritual aspect, Saturn is the ruler of structure as I've often said, it's the builder, a negative aspect is deterioration of the bone structure, causing it to go crusty, medical term could have been arthritis but he never told me he had his back x-rayed. Externally, in Dave's case there was much pressure put on him by his Father and he was always called upon to move things or do things for his father, which would suddenly cause him to have to leave and go home. His father had been very active but, because he had his vertebrae fused he couldn't do things himself, and so in my analysis the external affected area had caused the internal reaction.

The same philosophy I'd use for a person dying of cancer, it's not the cancer that is killing them, but it still needs to be treated, even when consciously becoming aware of the situation. In talking of Cancer the Cancerian sign is one of concern and worry, it's an active sign, cardinal. It's also seen with the Moon and

4th house, and it's where we harbour memories. When the Moon transits the constellation it will tell us how we will emotionally react.

Referring back to Wayne's Natal Chart his Moon was in Leo, in the 3rd house but, because it is several years ago, I'm not sure if the time was right but the opposites/square and quincunx's will be right. And the formula I created worked. In my notes I put alternate foods and ways of preparing them.

Looking at the flow chart there is only 1° orb in the square between Moon to Mars. Mooin Leo meant his emotions are very proud, but the work environment could be dirty and grimy which would emotionally upset him, he had a square to working, this is not unusual many people go to work and hate but, there's an incentive for them to keep doing it. As to the field of work the chart shows the Moon makes a quincunx to Saturn so in any type of work he'd find his emotions interfering with his work, which could be the memories he has of his upbringing,

Below is his Flow Chart I've put in the house position but, as I've said I'm not confident. I couldn't get the same time that initially put planets into the houses. It's probably so simple I've just overlooked something.

W.R. Natal chart Converted to a Flow Chart for 1 Nov 1961

W.R. Flow Chart Combining Natal & Progressed														
PLANET	Degree '	Trip	Quad	Acid/Alk	**HOUSE**	Trip	Quad	Acid/Alk	**CONSTELLATION**	Trip	Quad	Acid/Alk	**ASPECTS**	Food
Jupiter	6 52	Fire	Mutable	Alkaline	5th	Fire	Fixed	Acid	Aquarius	Air	Fixed	Alkaline	Jup□ Sn	Animal/
Sun	8 05	Fire	Fixed	Alkaline	6th	Earth	Mutable	Alkaline	Scorpio	Water	Fixed	Acid		Liver
Pluto	9 41	Water	Fixed	Acid	3rd	Air	Mutable	Acid	Virgo	Earth	Mutable	Acid		
Neptune	10 56	Water	Mutable	Acid	6th	Earth	Mutable	Alkaline	Scorpio	Water	Fixed	Acid	Nep □ Mn	Fish/
Moon	12 14	Water	Cardinal	Acid	11th	Air	Fixed	Alkaline	Leo	Fire	Fixed	Alkaline		pd milk/
*Neptune	12 30	Water	Mutable	Acid	6th	Earth	Mutable	Acid	Scorpio	Water	Fixed	Acid		Oranges
Venus	12 30	Earth/Air	Cardinal	Alk/Acid	2nd	Earth	Fixed	Acid	Sagittarius	Fire	Fixed	Alkaline		
Venus	16 55	Earth/Air	Fix/Card	Acid/Alk	5th	Fire	Fixed	Alkaline	Libra	Air	Cardinal	Alkaline		
Nth Node	19 46R				11th	Air	Fixed	Alkaline	Leo	Fire	Fixed	Alkaline		
Mars	20 39	Fire	Cardinal	Alkaline	1st	Fire	Cardinal	Alkaline	Scorpio	Water	Fixed	Acid		Pepper
Mercury	22 10	Air/Earth	Mutable	Alk/Acid	3rd	Air	Mutable	Alkaline	Sagittarius	Fire	Mutable	Alkaline		
Mercury	22 19	Air/Earth	Card/Mut	Alk/Acid	1st	Fire	Cardinal	Alkaline	Libra	Air	Cardinal	Alkaline	Merc □ Sat	Celery
Mars	22 47	Fire	Cardinal	Alkaline	3rd	Air	Mutable	Alkaline	Sagittarius	Fire	Mutable	Alkaline		
Sun	22 52	Fire	Fixed	Alkaline	3rd	Air	Mutable	Alkaline	Sagittarius	Fire	Mutable	Alkaline		
Nth Node	24 11				11th	Air	Fixed	Alkaline	Leo	Fire	Fixed	Alkaline	NN ⚻ Sat	
Saturn	24 11	Earth	Cardinal	Acid	4th	Water	Cardinal	Acid	Capricorn	Earth	Cardinal	Acid	Leo ⚻ Cap	Oil
Moon	26 10	Water	Cardinal	Acid	6th	Earth	Mutable	Acid	Pisces	Water	Mutable	Acid		
Saturn	27 49	Earth	Cardinal	Acid	4th	Water	Cardinal	Acid	Capricorn	Earth	Cardinal	Acid		
Jupiter	29 06	Fire	Mutable	Alkaline	4th	Water	Cardinal	Acid	Capricorn	Earth	Cardinal	Acid	Jup ⚻ Ur	Liver
Uranus	29 58	Air	Fixed	Alkaline	11th	Air	Fixed	Alkaline	Virgo	Earth	Mutable	Acid	Ur □ Chir	Salt
Uranus	00 31R	Air	Fixed	Alkaline	11th	Air	Fixed	Alkaline	Virgo	Earth	Mutable	Acid		
Chiron	01 57	Not Enough Information			9th	Fire	Mutable	Alkaline	Pisces	Water	Mutable	Acid		

Green is the progressed planets

With Wayne I took a different approach to other formulas. I could clearly see the skin cancer and so I focused on his arms; immediately looking in his chart for Mercury/Jupiter and Saturn/Capricorn aspects; and, of course Pluto. As you can see Pluto is only 1 degree from the Sun. The aspects to constellations is not harsh. But, there is a 210 deg made between Aquarius & Virgo the alternate partner to a quincunx.

Progressed Jupiter is in Aquarius squaring Sun and Pluto; Mercury square Saturn; Natal Jupiter quincunx's Uranus. In focusing on these aspects we see Progressed Jupiter in Aquarius Square Sun in Scorpio. This means he'd have problems with weight/problems with liver, in fact he was extremely slim. It means his hopes probably would have been adverse to his fathers. It indicates to combine in the formula- air dried liver with Oranges and salt

Mercury in Libra Square Saturn in Capricorn
Mercury has to do with arms, breathing, brain, the nervous system and air, being in Libra means there would be a doubling of air and a sweetness; being square means the air would be blocked by the influence of Saturn, and there would be breathing problems. Asthma/Emphysema, his breathing was stressed at the time I wasn't focusing on lung illnesses, but of course it was indicated with skin cancer on the arms, without realising it when doing formulas, you can be helping other problem. But, Saturn in Capricorn gives a double influence. Capricorn/Saturn element earth has a gravely/soil effect.

Saturn is 5 degrees away from Jupiter, this means he'd be quite joyful when working, but almost immediately he gets started he'd be called away, with the Uranus 150 deg. aspect, this wouldn't be good for permanent work. With Saturn we get a crust when aspects affect it. Saturn with Jupiter would normally be productive, but Uranus which rules wild animals up until 2 months earlier he'd been helping to look after wildlife at home. it's made a health aspect

What I'm going to do unlike previous explanations I'm going to write the formula and show the influence.

<u>Formula</u>
Cow Liver to charcoal – Jupiter rules the liver
Olive Oil – Olives are said to be a Sun plant, oil is Saturn
Salt – is required for Uranian problems
Water – Element of Scorpio
Orange Juice – A Sun fruit
Previous made Pepper Celery cooked and dried – Celery – Mercury and Mars is pepper
Hoki dried - Fish is symbol of Neptune

I've omitted quantities because they change and are individual.

Capricorn also refers to coal/charcoal
Oranges are excellent for Magnesium a product required for the heart and the Sun is round and Orange.

Liver is a product of Jupiter for skin, I possibly could have used chicken livers they would have combined both Mercury and Jupiter and cause the formula to have more coverage.
Salt is a product of Uranus

When Jupiter is in Capricorn we are looking for something that renews itself or expands, we never think of the skin renewing itself but if we have a cut it eventually sows itself up in a healthy person. We often hear of the liver growing when parts have been cut off, these are the two functioning rulerships of Jupiter. Jupiter being in Capricorn means we want something that will eventually form a structure or is active, its element is earth Jupiter's the planet that rules philosophies/religions and law. An obvious food group of Jupiter is by now shouldn't be surprising is liver which can be bought in any supermarket I was torn between 3 food groups liver/ bacon chips & pork, I used liver because, I've never found it to fail, another quality I was looking for was earth substance required by Capricorn and earth is created by the drying substances, and because Jupiter has a fire element, this led me to put the liver under the grill to dry it out. The next step was to find not the element of Capricorn, but a material that had been doubly processed, oil has been contributed to Capricorn. The theory behind it is that as the earth ages it creates oil so to the formula I add 1/4 cup of Olive Oil with also a characteristic of Jupiter where a little bit goes a long way.

Where Mars is concerned there was a formula I had fallen back onto which I had made and used earlier. It was simple and affective I put pepper with celery and dried the ingredients in the microwave. Celery is an air product which is Gemini, I probably would have preferred comfrey but it's very hard to find and for myself harder to grow, why because the description of the buds of the comfrey plant are described as a hand usually having 5 flowers and the skin cancer affected his hands and arms. The Asians believe it has many beneficial qualities, and I've read so has celery but not as much potassium, in saying that it has a high degree of sodium mixing it with pepper added the potassium where combined gives a Mars/Mercury influence, whilst there is little carbohydrates in celery I've read and interestingly on the internet under 'what is celery high in' that the calories in the celery comes from sugars, but, the good qualities far outweigh the bad. But, there again in using celery it added the natural sugars I required for Mercury being in Libra.

The Moon and Neptune square led me to add Hoki they feed on small crustaceans which is ruled by Saturn it's found in the oceans which is ruled by Scorpio/Pluto which relates to the Neptunian and Sun influence we see in Scorpio. The process I took was to cook it in a little bit of water let it dry out and combined the below ingredient to rub into his chest and ankles.

Over 15 years I've worked on many areas the above have been the more intense. Before I go on I should mention that my own health history has constantly changed which I mentioned in the early part of this book, when I've concentrated on my own health I've seemed to have problem areas disappear i.e. diabetes which didn't show up in my last blood test which has amazed doctors, on rare occasions I'd inform them that I was working on my own health formulas. The following outline those minor areas I've worked on for the many people who have passed through my life:

Acute Anaemia; Continual headaches; Back pain; Pain through back fusion; Smokers cough; Vaginal itching; Hives; Sore throat; Boils; Broken toe pain; a neighbour who had bone cancer and a reaction to tablets, when his family thought he was looking better they took him away; Goitre; Leukaemia with friend; Severe neck pain.

DAVE S. TREATMENT FOR BACK PAIN

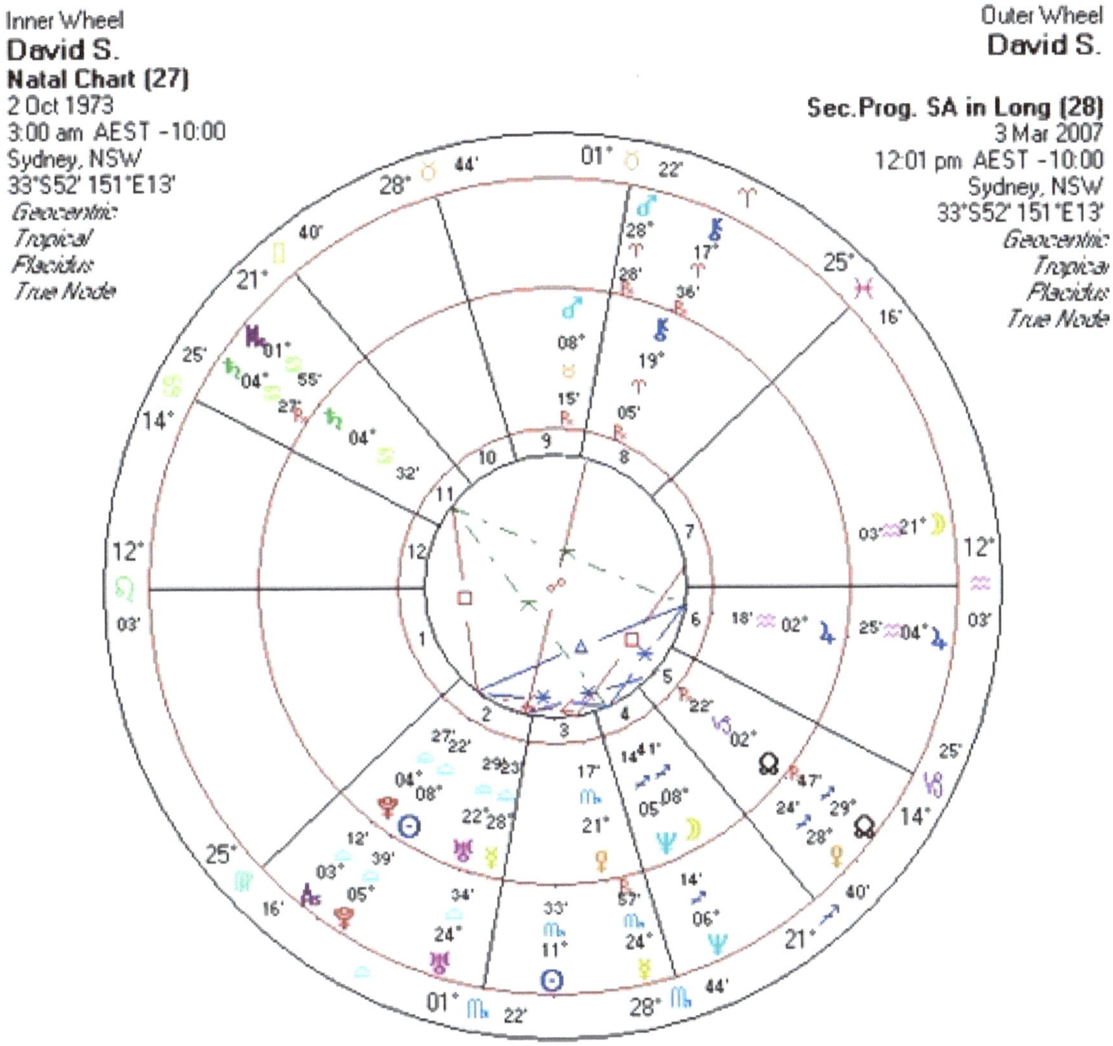

CHART 13

Above is David's Natal Chart (Inner Wheel) and Progressed Chart (Outer Wheel). Chart 10 shows the aspects he was born with, the red line show considerable squares, whilst he had the ability to overlook them with a cheerful outer appearance when with others, and there was an easy flow with his family, as long as he didn't work. It also shows that he was in tune mentally with his father, but physically not able to cope and that the father would put the pressure on him with Saturn being prior to the Sun which is the physical being. There was an obvious opposition with Uranus Mars and Mercury and for some reason going through my notes I didn't put it in but, I did add a transiting aspect of Uranus in Pisces. And in the formula there is a food for Mars which is pepper and Mercury which is celery.

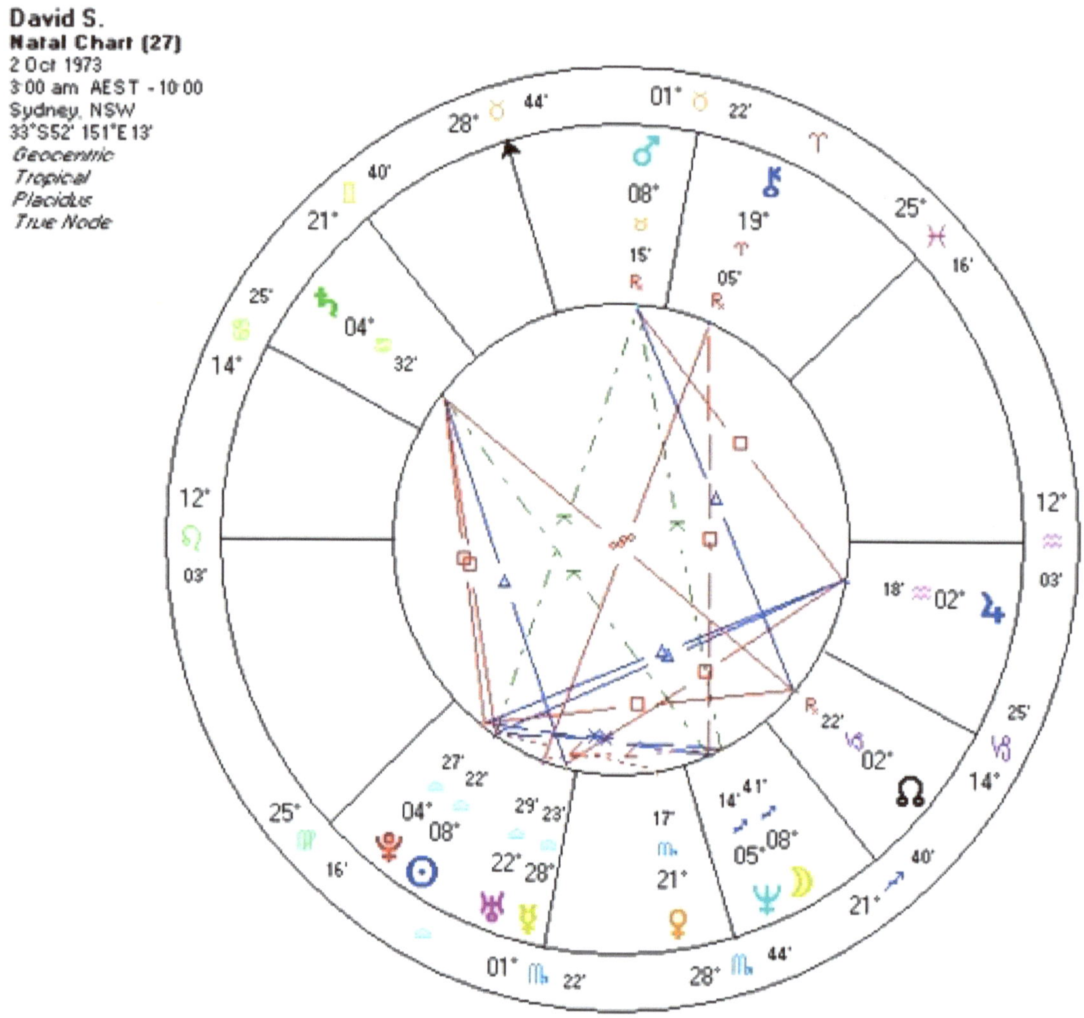

CHART 14

Pain is evident with Saturn Square Pluto and the Sun. There is also Mars making 150 deg. Angle to the Sun.

Below is David's Flow Chart putting planets in order of degrees helps to know which order one prepares the food. When there are conjuncts you combine; aspects up to 10 deg. away you prepare separately; sometimes I've done 3 separate formulas, with N.R. I was preparing formulas on the spot in particular making massage oils apart from acupuncture and hypnosis.

The foods I chose were those in the flow chart, the final product was rubbed into his back. I've noticed the Uranus influence missing were I did add Mushrooms this carries water and unpredictably grows in moist weather.

The foods I used took into consideration the Planet House and Constellation e.g. Jupiter/Liver, Saturn/Mylanta or Charcoal found in burnt wood, Cancer/Milk, Neptune Wine. In my notes I haven't put the exact time which makes the house positions dubious. Whilst the formula worked, on what I did, but obviously I could have had the correct time.

D.S. Natal & Progressed Flow Chart based on DOB & 3 March 2007

PLANET	Degree '	Trip	Quad	Acid/Alk	HOUSE	Trip	Quad	Acid/Alk	CONSTELLATION	Trip	Quad	Acid/Alk	ASPECTS	Food
Jupiter	2 18	Fire	Mutable	Alk	**6th**	Earth	Mutable	Acid	Aquarius	Air	Fixed	Alk		
Jupiter	4 25	Fire	Mutable	Alk	6th	Earth	Mutable	Acid	Aquarius	Air	Fixed	Alk	**Jup/Sat**	**Cow Liver**
Pluto‡	4 27	Water	Fixed	Acid	**2nd**	Earth	Fixed	Acid	**Libra**	Air	Cardinal	Alk	Jup/Sun	Mylanta
Saturn	4 27 R	Earth	Cardinal	Acid	11th	Air	Fixed	Alk	Cancer	Water	Cardinal	Acid		**Pd Milk**
Saturn	4 32	Earth	Cardinal	Acid	**11th**	Air	Fixed	Acid	**Cancer**	Water	Cardinal	Acid	**Sat/Nep**	Wine + Oil
														Egg Shell
Neptune‡	5 14	Water	Mutable	Acid	**4th**	Water	Cardinal	Acid	**Sagittarius**	Fire	Mutable	Alk		**Kelp**
Pluto‡	5 39	Water	Fixed	Acid	2nd	Earth	Fixed	Acid	Libra	Air	Cardinal	Alk		Red Fish
Mars	8 15 R	Fire	Cardinal	Alk	**9th**	**Fire**	Mutable	Alk	**Taurus**	Earth	Fixed	Acid	**Mar/Sun**	**Cow Necks**
Sun	8 22	Fire	Fixed	Alk	**2nd**	Earth	Fixed	Acid	**Libra**	Air	Cardinal	Alk		Pepper Grain
Moon	8 41	Water	Cardinal	Acid	**4th**	Water	Cardinal	Acid	**Sagittarius**	Fire	Mutable	Alk		Oranges
Sun	11 33	Fire	Fixed	Alk	3rd	Air	Mutable	Alk	Scorpio	Water	Fixed	Acid	Sun/Mn	
Chiron	17 36	Insufficient Information			**8th**	Water	Fixed	Acid	**Aries**	Fire	Cardinal	Alk		
Moon	21 03	Water	Cardinal	Acid	6th	Earth	Mutable	Acid	Aquarius	Air	Fixed	Alk	Mn/Vn	
Venus	21 17	E/Air	Fix/Card	Acid/Alk	**3rd**	Air	Mutable	Alk	**Scorpio**	Water	Fixed	Acid		
Uranus‡	22/24	Air	Fixed	Alk	**2nd**	Earth	Fixed	Acid	**Libra**	Air	Cardinal	Alk		
Mercury	24 57	Air/Earth	Mutable	Acid/Alk	3rd	Air	Mutable	Alk	Scorpio	Water	Fixed	Acid		
Mercury	28 23	Air/Earth	Mutable	Acid/Alk	**2nd**	Earth	Fixed	Acid	**Libra**	Air	Cardinal	Alk	**Merc/NN**	Celery
Venus	28 24	E/Air	Fix/Card	Acid/Alk	5th	Fire	Fixed	Alk	Sagittarius	Fire	Mutable	Alk		Cod Liver
Nth Node	29 47				5th	Fire	Fixed	Alk	Sagittarius	Fire	Mutable	Alk		Oil
Nth Node	2 22				**5th**	Fire	Fixed	Alk	**Capricorn**	Earth	Cardinal	Acid		

N.S. – GIRL WITH HORMONAL PROBLEMS NATAL & PROGRESSED CHART

CHART 15

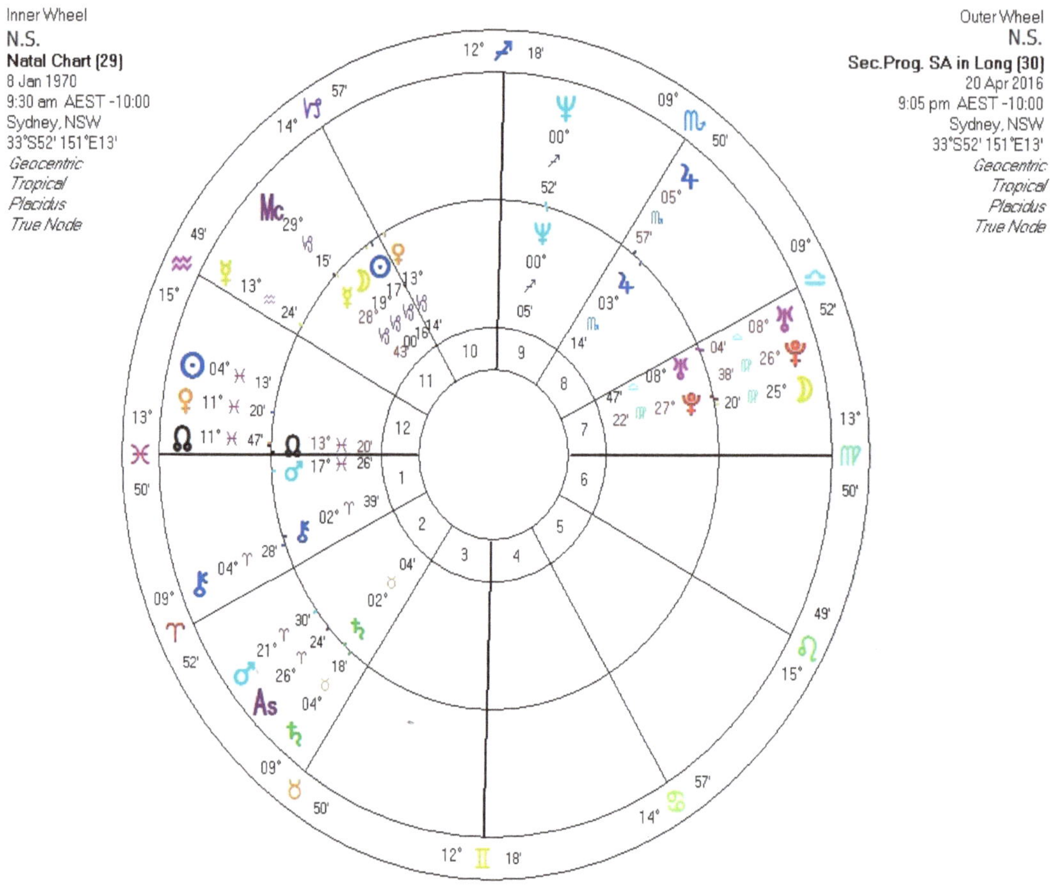

N. S. – GIRL WITH HORMONAL PROBLEMS

Above, with the first Astrology Chart I have combined the natal and progress planetary aspects for Nicky S for this current time, it is a sudden impulse to include her chart it explains why I couldn't do a formula. Nicky and I, became friends just over 10 years ago and whilst I've given her a formula I'd created, I found she kept using something the Doctor would give her to trial and for a long while it was never used. It came a time when she found the prescription was just too expensive. When we met I had opened a counselling business, but I've found working for myself is just untimely as my chart shows, so having my mind on creating a full formula was too stressful where there is no money coming in. When you're spending money you're always trying to think how to make the business work. And whilst I had included Astrology I was looking into law and legal forms for the public.

Nicky and I only knew each other for about 6 months when periodically she had what appeared to be menstrual problems. Menstrual problems for women are just natural; and one expects discomfort; but, change of life at 36 is a bit early.

I'd remember she told me that she was waking up in cold sweats drenched. Which I believe has continues till this day.

I remember starting a few times to do a formula for her, but I felt I wasn't in the right mind space. This is important, whilst a flippant remark, the mind is ruled by Mercury and Mercury rules Gemini and Virgo. As a consequence, I said to her 'I mightn't be able to do a formula for you'. In what I said I realise now, it all had to do with timing, and in reflecting in her progressed chart for 2006. I can clearly see now my natal Nth Node is conjuncting her Nth Node in Pisces with a 4° difference. And being opposite to Virgo means opposite to formulas. Pisces has to do with instinct and alcohol, and there were many a memorable times and still are when I see her. But what comes easy is opposite to being practical and looking into health issues. As such my comment was instinctual and, correct.

V.S. Natal & Progressed Flow Chart based on DOB & 20 April 2016

PLANET	Degree	Trip	Quad	Alk/Acid	HOUSE	Trip	Quad	Alk/Acid	CONSTELLATION	Trip	Quad	Alk/Acid	Aspect	Food
Sun	04 13	Fire	Fixed	Alk	11th	Air	Fixed	Alk	Pisces	Water	Mutable	Acid	Pl/Sun	Calc Sulph
Saturn	04 18	Earth	Cardinal	Acid	2nd	Earth	Fixed	Acid	Taurus	Earth	Fixed	Acid		& Magnesium
Chiron	04 28				1st	Fire	Cardinal	Alk	Aries	Fire	Cardinal	Alk		
Jupiter	05 57	Fire	Mutable	Alk	8th	Water	Fixed	Acid	Scorpio	Water	Fixed	Acid		
Uranus	08 04 R	Air	Fixed	Alk	7th	Air	Cardinal	Alk	Libra	Air	Cardinal	Alk	Sun/Ur	
Uranus‡	08 47	Air	Fixed	Alk	**7th**	Air	Cardinal	Alk	**Libra**	Air	Cardinal	Alk	Ur/Venus	Salt/Sugar
Venus	11 20	Ea/Air	Fix/Card	Ac/Alk	12th	Water	Mutable	Acid	Pisces	Water	Mutable	Acid		
Nth Node‡	11 /13 deg				**12th**	Water	Mutable	Acid	**Pisces**	Water	Mutable	Acid		
Venus	13 14	Ea/Air	Fix/Card	Ac/Alk	**10th**	Earth	Cardinal	Acid	**Capricorn**	Earth	Cardinal	Acid		
Mercury	13 24	Air/Ea	Mut/Mut	Alk/Acid	11th	Air	Fixed	Alk	Aquarius	Air	Fixed	Alk		
Sun	17 16	Fire	Fixed	Alk	**11th**	Air	Fixed	Alk	**Capricorn**	Earth	Cardinal	Acid	Sun/Mars	
Mars	17 26	Fire	Cardinal	Alk	**1st**	Fire	Cardinal	Alk	**Pisces**	Water	Mutable	Acid		pd milk/water
Moon	19 00	Water	Cardinal	Acid	**11th**	Air	Fixed	Alk	**Capricorn**	Earth	Cardinal	Acid	Mn/Mars	Chilli sauce
Mars	21 30	Fire	Cardinal	Alk	2nd	Earth	Fixed	Acid	Aries	Fire	Cardinal	Alk	Mars/Mn	Orange Pepper
Moon	25 20	Water	Cardinal	Acid	7th	Air	Cardinal	Alk	Virgo	Earth	Mutable	Acid		
Pluto ‡	26/27 deg	Water	Fixed	Acid	7th	Air	Cardinal	Alk	Virgo	Earth	Mutable	Acid	Pl/Nept	Fermented
Mercury	28 43	Air/Ea	Mut/Mut	Alk/Acid	**11th**	Air	Fixed	Alk	**Capricorn**	Earth	Cardinal	Acid		Hoki & Kelp
Neptune	00 05 /52'	Water	Mutable	Acid	**8th**	Water	Fixed	Acid	**Sagittarius**	Fire	Mutable	Alk	Nept/Sat	Grilled Liver
Saturn	02 04	Earth	Cardinal	Acid	**2nd**	Earth	Fixed	Acid	**Taurus**	Earth	Fixed	Acid	Sat/Jup	Gin-Charcoal
Jupiter	03 14	Fire	Mutable	Alk	**8th**	Water	Fixed	Acid	**Scorpio**	Water	Fixed	Acid		Horse Fertisliser

Later in 2011 when I moved to Wollongong I tried again this time I was able to do a partial formula, where she informed me it worked, but she had discontinued using it because when the symptoms came on, it was in the middle of the night and she couldn't bring herself to rub a fine granule formula on herself. I remember it had to do with seaweed. And, looking at the progressed chart for 2006 below it shows North Node, Sun and Venus in Pisces 12th house relating intensively to sleep. This means she was prone to being very tired, and maybe suffering from anaemia and a lazy thyroid; and, if I remember correctly she did tell me doctors thought there were thyroid problems. Pisces planetary ruler is Neptune ruler of the sea and so Seaweed is very desired having an iodine content.

I found out recently that she'd been sick for the last 2 weeks with serious back pain with extremely heavy period, and for many years she hadn't slept in her own bed. This time I've found my mind is not so subjective meaning full of personal feeling, BECAUSE I realize we didn't only have our Nth Nodes conjunct, my moon is also conjuncting the Nth Nodes, which makes an easy flow to Jupiter in Aries in 8th house (8th house is everything or nothing). Which is opposite my Natal Saturn. And, I've found work brings no joy. It reminds me of how it used to upset my daughter in particular when I worked from home, and this is an obvious outcome with Saturn making a 150° to Moon & North Node.

With Nicky's chart going purely by the symptoms you'd expect to see the planet that gives pain, Saturn involved, (write these down) Saturn also rules the vertebrae of the body. For the pain to be so severe for having to take 2 weeks off work, it meant something would have emphasised it. When we look at direct oppositions we see Saturn conjunct Saturn in Taurus, Taurus is also the ruler of the 2nd house where these two planets are, as such we have two planets in Taurus both in the 2nd house which gives an emphasis of 4 and probably giving an overactive thyroid. This causes gravel, being opposite the natal and progressed Jupiter that rules the Liver means Saturn would be stopping the necessary hormone for the liver to function properly, thyroxine. This is because Thyroid is ruled by Taurus. This. aspect would also cause a problem with her periods because of Saturn in Taurus opposite Scorpio.

Natal and Progressed Jupiter are both in Scorpio, Scorpio rules the Reproductive Organs and Jupiter rules excess. Jupiter is the planet that rules Sagittarius which are known for their extravagance and would be the cause of the heavy and excessive bleeding. When Jupiter is in Scorpio, and 8th house, (normally ruled by Scorpio) – I'd be confident in using the oil from horse manure, Sagittarius symbol being the Centaur half man half horse, 8th house rules manure.

This is only a suggested part to a formula, and I was in shock when I first thought of trying it. Of course I wouldn't include it without permission, and when I first used it with Wayne, he praised me immediately, this came after the other formulas I've written up on, and had an instant effect on allowing him to breath. Where in truth I needed to do extra formulas on other planetary aspects I neglected. Returning to Nicky, it shows that Jupiter makes a positive aspect to the Sun if Saturn hadn't been causing so much pain.

Aspects of Nicky haven't moved much in 10 years and can be seen in both charts 14 and 15, one has to realise that the planets for a progressed chart will have only moved 46 days. (Refer Planets Movements for more information). In saying that the Moon has moved the furthest, where it was in Taurus and now in Virgo.

When North Node is in Pisces it also accompanies fear. Fear that things can't be done. It makes things either very confused or very spiritual. Because Nicky and I had formed a deep bond it made me fearful that I'd miss something and fail. With Nicky's North Nodes they give the ability of perception because of the natural flow to the Sun and a number of planets. She has a beauty which is noticed with Venus conjunct. But, she often says to people who compliment her that it's not external beauty that counts. This straight forwardness comes from the Moon square Mars, where there is a double Mars influence with Mars in Aries, in 2006 Mars directness flowed on to the male form, and it made her action more direct, now as the saying goes which I haven't checked with her she might find men contrary and males acting out on her, working in an aged care centre. In addition, the North Node and Venus are in Pisces, this means she wouldn't see herself as others do, but it's also accompanied by her Natal Mercury is in Capricorn which brings about focus.

Where I first mentioned Nicky having Night sweats and waking up soaking wet the first thing the Doctors thought was that it related to menopause, a time when a woman has irregular periods and where the blood flow can be erratic. But, in fact night sweats relate to the planet Mars and constellation Aries. When it was first happening Natal Mars is in Pisces in its own 1st house. This would mean excessive sweating whilst asleep when Mars is affected. But it becomes more obvious with Progressed Mars squaring the Sun. Mars rules the 'red' blood cells. In the flow chart I've suggested Orange Peppers.

Unfortunately, I haven't the facility to make the formula up, and these aspects for some are constant because of the progressed planets, or because of the persons own natal chart, but what influenced the illness were the transiting planets, where her illness lasted a fortnight. Progressed signs can take 4 to 5 years to get over, and if there is another planet close another type of illness will show up. I haven't done enough research to know how long my formulas work only for Neville who had a stroke, and returned to limping after 15 months when his doctor couldn't get rid of a lump in his good leg.

Because I've omitted the Transits, I've included chart 16 for the 5 April 2016. Where the red lines show the number of planets affecting us at that time. The other unfortunate thing for myself is that Astrology is discriminated against and if a patient is unwilling to trial the formula, it would be necessary to talk to a doctor with an open mind and to get over prejudices, and this illusion that god didn't give us a path.

I've given the foods in the Flow Chart and the preparation is in the elements. The symbol of Constellations has already given us the primary food to work from. Whilst not required in Nicky's chart, with Gemini and Libra I use the anatomy part it rules i.e. sheep/cow brains; for Libra is cow's Kidney.

Alternatives are given under Foods of Constellations, for Libra problems relating to Kidneys, lately and I feel so stupid in saying this, I've found bi-carbonated soda should be looked into for those on dialysis, but in very small amounts in water. Soda is a cell salt for Libra and its planet Venus has 96.5% Carbon Dioxide, and in giving it to a father whose young child was in hospital for kidney problems, she was out in a few days instead of the weeks they thought she'd be in there.

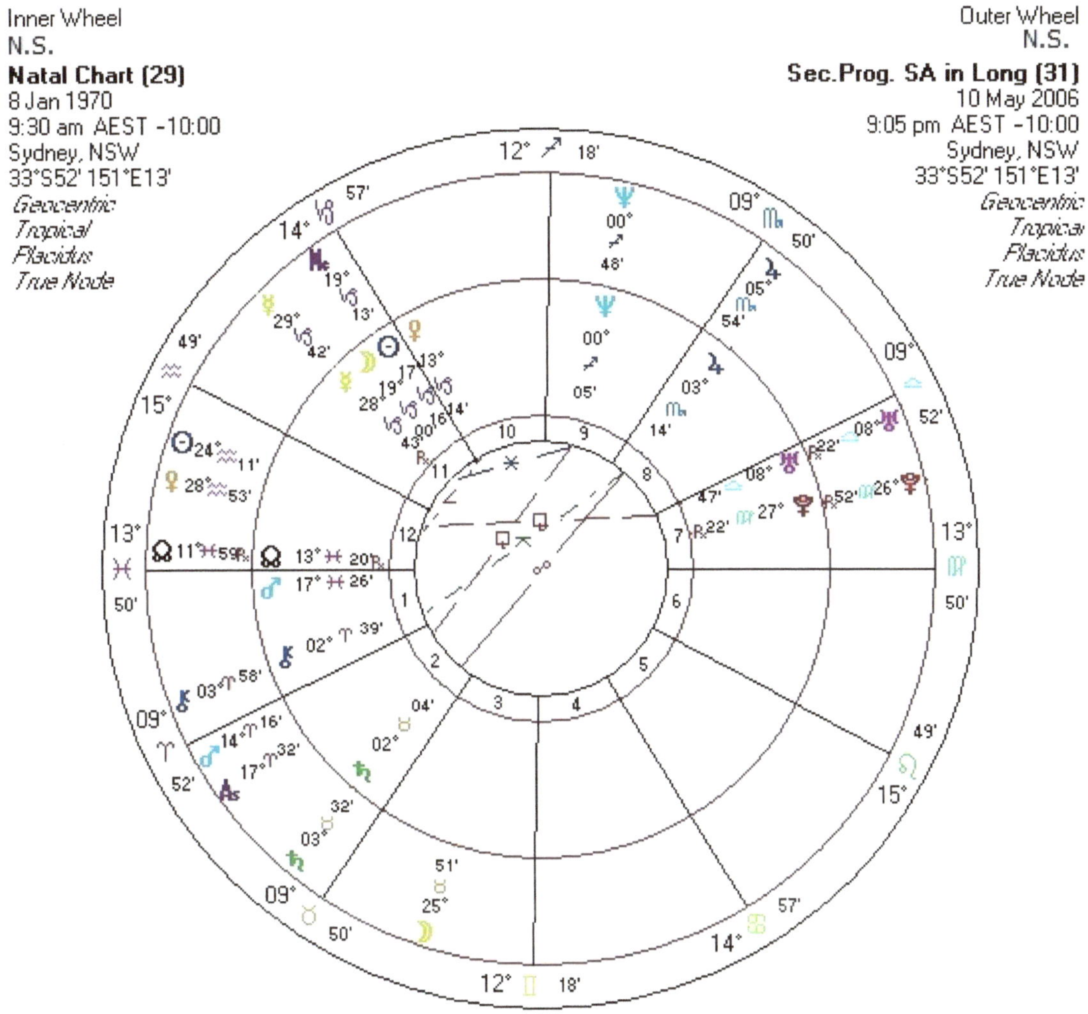

Chart 16

Chapter 12 - Calculating Your Chart Manually

The following works of Ursula Lewis who wrote 'Chart our Own Horoscope', is one of the best examples I've ever come across giving full examples which you can understand by following the steps which are outlined in text as well as providing illustrations You'll find it handy if you wish to sit the American or Australian Federation of Astrologer's FAA exam . Though it's not in its entirety the later part of this chapter gives the calculations for understanding the planets movements. I am putting this first because if you're at all sceptical you'll realise you'll need brains as well as beauty to go on. Except for those who've got computers :)

Text books required are: -

1. A Table of Houses
2. A book on Longitude and Latitude of the World
3. The Prime Meridians
4. A book showing Time Changes
5. Ephemeris giving sidereal time and planetary positions for a given date.
6. Table of proportional logarithms you will require in particular Table 33, Appendix A
7. A blank birth chart.

To calculate a chart, we begin with Date of Birth, Time of Birth and Place of Birth, it doesn't need to be a person it could be a country or event and we continue as follows:

Abbrev: L.M.T. > Local Mean Time; E.G.M.T. > Equivalent Greenwich Mean Time; ST. T. > Standard Time

Date: 7 November, 1972. Time: 2:24 pm ST.T. (Standard Time). Place: Los Angeles, Ca.

Longitude	Latitude	L.M.T	E.G.M.T.
118° W15'	34° N03'	+7m 00s	+7h 53m 00s

(For conversion for Longitude refer The Prime Meridians)
(E.G.M.T. found in the Ephemeris)
Steps:

	Hours	Minutes	Seconds
1. Clock time at p.m. birth	02	24	00 ST. T.
2. Local Mean Time is a variation from birth time and gives >	+	7	00
True Local Time =	02	31	00
3. 10 Second Interval Correction	+		26
This is a calculation on hr & mins 1 Hr = 10s 1-6 min = 1s	02	31	26
4. Sidereal Time of the 7th as birth	+15	07	08
occurred after 12 Noon	17	38	34
5. Longitudinal Correction (Formula: Multiply by 2, divide result by 3, Always Add result when birth place was west of Greenwich; deduct result when birth place is east of Greenwich)	+00	01	19
The final calculated Sidereal Time =	17	39	53

Brought Forward	Hour	Minute	Seconds
The final calculated Sidereal Time =	17	39	53

This figure is used in the Table of Houses along with the Latitude to find the Ascendant, Medium Coeli M.C., Nadir, Descendent & other 8 house positions.

6. Refer Step 2 the True Local

Time	02	31	00
Refer E.G.M.T.	+07	53	00
	09	94	00

The 6th Step is used to find planetary movements by converting the h/m/s into a logarithm the 9h 94m Equivalent = 10hrs 24m 00s

This is the Greenwich Mean Time on the 7th Nov.

The C.L. (Constant Log) for 10hr 24m = .3632 (This is the planetary correction between 7th and 8th November)

Equations

The earth is divided into twenty-four prime meridians, each one 15° distant from the next. This is longitude calculation. From Greenwich Observatory in England, which has been designated as 0°. (See The Prime Meridian Table). In all places clocks are synchronized to the standard time in that prime meridian. This system has the obvious advantage of standardization, but it does not assure true time. Unless a person was born exactly on a prime meridian, the true time of his birth is not the same as the official time. Hence the official time has to be converted to True Local Time.

To return to our example: The clocks for Los Angeles are set for the 120th prime meridian, but since Los Angeles is only 118° N15' it is actually 1° 45' east of its prime meridian or true time centre. Now, the Sun travels 1° every 4 minutes, and 15° x 4 minutes equals 60 minutes, which equals 1 hour. This is why each of the prime meridians are 15° for 1 hour apart. Since Los Angeles is short by 1° 45' of the nearest prime meridian, and 1° is equivalent to 4 minutes of the time and 45' is equivalent to 3 minutes of time, the difference between the True Local Time and the official time is Los Angeles is 7 minutes. The 7 minutes are added to the official time. The 7 minutes are added to the official time because Los Angeles is east of its true time centre. (If the birthplace is west of the prime meridian, then the time difference is subtracted.)

In the Birth Calculation, Step 1 the clock time is noted as 2 hours, 24 minutes, 00 seconds. Since Daylight Saving Time was not in effect in November 1972 it is Marked ST. T. which stands for Standard Time. At this point let me note that we are working with a *noontime ephemeris* in which each day begins at Greenwich Noon Time and ends at noon the next day. From 12 Noon 7 November 1972, 2 hours, 24 minutes and 00 seconds have passed in our example case.

Step 2 concerns the Local Mean Time variation in this case the adjustment is 7 minutes, 00 seconds, and it is added to the clock time.
Step 3 is a 10-second interval correction. Here we begin to adjust for Sidereal Time. ***Sidereal Time progresses by approximately 4 minutes per day.***

The 4 minutes-per-day accumulation time 4 years equals just slightly over 24 hours, or one extra day – which is inserted into our calendar in February 29 every four years. Sidereal Time is the most accurate method of time reckoning and is used in every astronomical observatory.

Since the ephemeris gives only one Sidereal Time per day –with all ephemeris usually an adjustment has to be made if the birth time was not exactly noon or midnight, in this case noon. The calculations in the

above example has a correction of 26 seconds, which is obtained from **The Prime Meridians**, which I included for your convenience. By checking the Prime Meridians, you will note that Two hours require a correction of 20 seconds, and 31 minutes require a correction of 6 seconds. The 10-second interval correction of 26 seconds is added to the L.M.T of 2 hours, 31 minutes, 00 seconds for a total of 2 hours, 31 minutes, 26 seconds (Step 1 plus 2, plus 3).

For **Step 4** we have to consult *Raphael's Ephemeris,* November 1972. The ephemeris page for 7 November shows the Sidereal Time as 15 hours, 07 minutes 08 seconds. We add this Sidereal Time to Step 3, which gives a total of 17 hours, 38 Minutes, 34 seconds.

Step 5 is the *longitudinal correction.* The rule for this is s follows. Longitude times 2 divided by 3. The longitude of Los Angeles is 118° W15' . Multiply 118 by 2, which gives 236. Divide this figure by 3 and the result is 78 with 2 remaining. Since the remainder is more than half the divisor, the total is rounded off to 79. This correction is always in seconds; 79 seconds would be 1 minute and 19 seconds. The longitudinal correction is *added to the Sidereal Time:* the result in the Calculations above is 17 hours, 39 minutes, 53 seconds – the Final Calculated True Local Sidereal Time.

Notice that we dropped the 15' in the 118° W 15' longitude figure. Whenever the minute figure is less than 30, it is dropped. If it is more than 30, the degree figure is increased by 1. Also note that the longitudinal correction is *added if the birthplace was west* of Greenwich but *deducted if the birthplace was east* of Greenwich. With the completion of Step 5, we are ready to set up the horoscope wheel.

TABLE OF HOUSES

TABLE OF HOUSES FOR LATITUDES 20° to 56°

UPPER MERIDIAN, CUSP OF 10th H.

	H.	M.	S.			H.	M.	S.			H.	M.	S.
SID. T.	17	33	51}	♐		17	38	13}	♐	25°	17	42	34}
ARC	263°	27'.8	}	24°		264°	33'. 1 }				265°	38.	5' }
H.	11	12	1	2	3	11	12	1	2	3	11	12	1
Lat.	♑	♒	♓	♈	♉	♑	♒	♓	♈	♓	♑	♒	♓
°	°	°	°	°	°	°	°	°	°	°	°	°	°
22	18	16.6	21 23	28.5	29	19.8	18	22.5	29.8	30	21	19	24 15
23	19	16.3	21 17	28.7	29	19.6	18	22 44	29.9	30	21	18.8	24 11
24	18	16.1	21 11	28.9	29	19.5	17	22 39	0.1 ♉	30	21	18.5	24 07
25	18	15.8	21 5	29.1	29	19.3	17	22 34	0.3	♊	20	18.3	24 03
26	18	15.5	20 59	29.3	29	19.2	17	22 29	0.5	0.2	20	0	23 59
27	18	15.3	20 53	29.5	29	0	17	22 24	0.7	0.3	0	17.8	23 55
28	18	0	20 46	29.7	29	18.8	16	22 18	1	0.5	20	17.5	23 50
29	18	14.7	20 39	29.9	30	18.7	16	22 12	1.2	0.7	20	17.2	23 45
30	17	14.4	20 31	0.1 ♉	30	18.5	16	22 06	1.4	0.8	20	16.9	23 40
31	17	14.1	20 24	0.3	30	18.3	15	21 59	1.6	1	19	16.6	23 35
32	17	13.8	20 16	0.5	.0 ♊	18.1	0	21 52	1.9	1.2	19	16.3	23 29
33	17	13.4	20 7	0.8	0.3	17.9	15	21 45	2.1	1.4	0	0	23 23
34	17	13.1	19 58	1	0.5	17.7	14	21 38	2.4	1.5	19	15.6	23 17
35	17	12.7	19 48	1.3	0.7	17.5	0	21 30	2.7	1.7	19	15.3	23 11
36	16	12.3	19 38	1.6	0.9	17.3	14	21 21	3	1.9	18	14.9	23 04
37	0	11.9	19 27	1.9	1.1	17.1	13	21 12	3.3	2.1	18	14.5	22 57
38	16	11.5	19 16	2.2	1.3	16	13	21 02	3.6	2.4	18	14.1	22 49

Step 2

Table 1

The Prime Meridians

Meridian	Hrs East of Greenwich
180°	-12 (dateline)
165°	-11
150°	-10
135°	-9
120°	-8
105°	-7
90°	-6
75°	-5
60°	-4
45°	-3
30°	-2
15°	-1
0° Greenwich England, Observatory	
	Hrs West of Greenwich
15°	+1
30°	+2
45°	+3
60°	+4
75°	+5
90°	+6
105°	+7
120°	+8
135°	+9
150+	+9
150°	+10
165°	+11
180°	+12 Dateline

Step 3

10 Second Interval Correction

hr	hr	min	sec
1	00	00	10
2	00	00	20
3	00	00	30
4	00	00	40
5	00	00	50
6	00	01	00
7	00	01	10
8	00	01	20
9	00	01	30
10	00	01	40
11	00	01	50
12	00	02	00
13	00	02	10
14	00	02	20
15	00	02	30
16	00	02	40
17	00	02	50
18	00	03	00
19	00	03	10
20	00	03	20
21	00	03	30
22	00	03	40
23	00	03	50
24	00	04	0
Minute	**H**	**M**	**S**
1-6	00	00	01
7-12	00	00	02
13-18	00	00	03
19-24	00	00	04
25-30	00	00	05
31-36	00	00	06
37-42	00	00	07
43-48	00	00	08
49-54	00	00	09
55-60	00	00	10

House Cusps

The House cusps are found by referring to the Table of Houses. In the calculations first applied in the 5th Step it shows the final calculation performed for the Sidereal Time which came to 17hrs 39m 53s all else under it doesn't apply to this. With this information we go to the Table of Houses which show the Latitude in the illustration we go to the page with Latitude that shows 22° to 56° the Latitude of Los Angeles is 34° N 03' in the Table it show the Sidereal Time closest to this, when you follow the Sidereal Time along as 17h 38m 13s this area also shows the 10th house as 25° Sagittarius (for this type of reference you need to know the symbols); when we go to the 5th row down we see the heading latitude. Being that Los Angeles latitude is 34° we follow the left hand column down to 34 and follow it across to the closest sidereal time as above, referring to the top of the column it shows 11 which refers to 11th house, then 12, 1, 2, 3 each being house cusps. When we run our figure down that 11th house is 17.7 to convert the figure after the decimal point we multiply the .7 by 6 which converts it to 42′. Here the result is 7° 42′. This is the simplest way of figuring out 7/10 of 1° or 60′. In the Table of Houses, the sign Capricorn appears under the 11th house and so the 11th house cusp will be 7° 42′ Capricorn.

Converting 12th House

In looking up the cusp of the 12th house in the Table of Houses we find 14.4. .4 is again multiplied by 6 giving us 24′ therefore 12th cusp is Aquarius 14° 24′. The Aquarian symbol is found under the cusp of the '12' house at the top of the column.

The 1st House

The first house is *never corrected.* It is copied exactly *as it appears.* In this case we find Pisces 21° 38′ as found in the Table of Houses.

The 2nd House

The cusp in the second house is given as 2.4. The 4 is again multiplied by 6 for a result of Taurus 2° 24′. If you cast your eye up the column of the 2nd house you will note there has been a sign change from Aries to Taurus. This can occur in other columns so you must always check.

The 3rd House

The cusp of the 3rd house is given as 1.5. The 5 is multiplied by 6 gives Gemini 1° 30′, notice again the sign change that occurs in the column.

The remaining 6 houses cusps are placed under opposing signs using the same degrees and minutes that have been calculated. For example 25° Sagittarius is on the 10th house, the house which opposes that is the 4th it's cusp will be 25° Gemini. 11th house cusp is 7° 42' Capricorn the opposing cusp will be the 5th house where the cusp will be 7° 42′ Cancer.

PROPORTIONAL LOGARITHMS

Minutes	Hours and Degrees												Minutes
	0	1	2	3	4	5	6	7	8	9	10	11	
0	3.1584	1.3802	1.0792	9031	7781	6812	6021	5351	4771	4260	3802	3388	0
1	3.1584	1.3730	1.0756	9007	7763	6798	6009	5341	4762	4252	3795	3382	1
2	2.8573	1.3660	1.0720	8983	7745	6784	5997	5330	4753	4244	3788	3375	2
3	2.6812	1.3590	1.0685	8959	7728	6769	5985	5320	4744	4236	3780	3368	3
4	2.5563	1.3522	1.0649	8935	7710	6755	5973	5310	4735	4228	3773	3362	4
5	2.4594	1.3454	1.0614	8912	7692	6741	5961	5300	4726	4220	3766	3355	5
6	2.3802	1.3388	1.0580	8888	7674	6726	5949	5289	4717	4212	3759	3349	6
7	2.3133	1.3323	1.0546	8865	7657	6712	5937	5279	4708	4204	3752	3342	7
8	2.2553	1.3258	1.0511	8842	7639	6698	5925	5269	4699	4196	3745	3336	8
9	2.2041	1.3195	1.0478	8819	7622	6684	5913	5259	4690	4188	3737	3329	9
10	2.1584	1.3133	1.0444	8796	7604	6670	5902	5249	4682	4180	3730	3323	10
11	2.1170	1.3071	1.0411	8773	7587	6656	5890	5239	4673	4172	3723	3316	11
12	2.0792	1.3010	1.0378	8751	7570	6642	5878	5229	4664	4164	3716	3310	12
13	2.0444	1.2950	1.0345	8728	7552	6628	5866	5219	4655	4156	3709	3303	13
14	2.0122	1.2891	1.0313	8706	7535	6614	5855	5209	4646	4149	3702	3297	14
15	1.9823	1.2833	1.0280	8683	7518	6600	5843	5199	4638	4141	3695	3291	15
16	1.9542	1.2775	1.0248	8661	7501	6587	5832	5189	4629	4133	3688	3284	16
17	1.9279	1.2719	1.0216	8639	7484	6573	5820	5179	4620	4125	3681	3278	17
18	1.9031	1.2663	1.0185	8617	7467	6559	5809	5169	4611	4117	3674	3271	18
19	1.8796	1.2607	1.0153	8595	7451	6546	5797	5159	4603	4109	3667	3265	19
20	1.8573	1.2553	1.0122	8573	7434	6532	5786	5149	4594	4102	3660	3258	20
21	1.8361	1.2499	1.0091	8552	7417	6519	5774	5139	4585	4094	3653	3252	21
22	1.8159	1.2445	1.0061	8530	7401	6505	5763	5129	4577	4086	3646	3246	22
23	1.7966	1.2393	1.0030	8509	7384	6492	5752	5120	4568	4079	3639	3239	23
24	1.7781	1.2341	1.0000	8487	7368	6478	5740	5110	4559	4071	3632	3233	24
25	1.7604	1.2289	0.9970	8466	7351	6465	5729	5100	4551	4063	3625	3227	25
26	1.7434	1.2239	0.9940	8445	7335	6451	5718	5090	4542	4055	3618	3220	26
27	1.7270	1.2188	0.9910	8424	7318	6438	5706	5081	4534	4048	3611	3214	27
28	1.7112	1.2139	0.9881	8403	7302	6425	5695	5071	4525	4040	3604	3208	28
29	1.6960	1.2090	0.9852	8382	7286	6412	5684	5061	4516	4032	3597	3201	29
30	1.6812	1.2041	0.9823	8361	7270	6398	5673	5051	4508	4025	3590	3195	30
31	1.6670	1.1993	0.9794	8341	7254	6385	5662	5042	4499	4017	3583	3189	31
32	1.6532	1.1946	0.9765	8320	7238	6372	5651	5032	4491	4010	3576	3183	32
33	1.6398	1.1899	0.9737	8300	7222	6359	5640	5023	4482	4002	3570	3176	33
34	1.6269	1.1852	0.9708	8279	7206	6346	5629	5013	4474	3995	3563	3170	34
35	1.6143	1.1806	0.9680	8259	7190	6333	5618	5003	4466	3987	3556	3164	35
36	1.6021	1.1761	0.9652	8239	7174	6320	5607	4994	4457	3979	3549	3157	36
37	1.5902	1.1716	0.9625	8219	7159	6307	5596	4984	4449	3972	3542	3151	37
38	1.5786	1.1671	0.9597	8199	7143	6294	5585	4975	4440	3964	3535	3145	38
39	1.5673	1.1627	0.9570	8179	7128	6282	5574	4965	4432	3957	3529	3139	39
40	1.5563	1.1584	0.9542	8159	7112	6269	5563	4956	4424	3949	3522	3133	40
41	1.5456	1.1540	0.9515	8140	7097	6256	5552	4947	4415	3942	3515	3126	41
42	1.5351	1.1498	0.9488	8120	7081	6243	5541	4937	4407	3934	3508	3120	42
43	1.5249	1.1455	0.9462	8101	7066	6231	5531	4928	4399	3927	3501	3114	43
44	1.5149	1.1413	0.9435	8081	7050	6218	5520	4918	4390	3919	3495	3108	44
45	1.5051	1.1372	0.9409	8062	7035	6205	5509	4909	4382	3912	3488	3102	45
46	1.4956	1.1331	0.9383	8043	7020	6193	5498	4900	4374	3905	3481	3096	46
47	1.4863	1.1290	0.9356	8023	7005	6180	5488	4890	4365	3897	3475	3089	47
48	1.4771	1.1249	0.9330	8004	6990	6168	5477	4881	4357	3890	3468	3083	48
49	1.4682	1.1209	0.9305	7985	6975	6155	5466	4872	4349	3882	3461	3077	49
50	1.4594	1.1170	0.9279	7966	6960	6143	5456	4863	4341	3875	3455	3071	50
51	1.4508	1.1130	0.9254	7947	6945	6131	5445	4853	4333	3868	3448	3065	51
52	1.4424	1.1091	0.9228	7929	6930	6118	5435	4844	4324	3860	3441	3059	52
53	1.4341	1.1053	0.9203	7910	6915	6106	5424	4835	4316	3853	3435	3053	53
54	1.4260	1.1015	0.9178	7891	6900	6094	5414	4826	4308	3846	3428	3047	54
55	1.4180	1.0977	0.9153	7873	6885	6081	5403	4817	4300	3838	3421	3041	55
56	1.4102	1.0939	0.9128	7854	6871	6069	5393	4808	4292	3831	3415	3035	56
57	1.4025	1.0902	0.9104	7836	6856	6057	5382	4798	4284	3824	3408	3028	57
58	1.3949	1.0865	0.9079	7818	6841	6045	5372	4789	4276	3817	3401	3022	58
59	1.3875	1.0828	0.9055	7800	6827	6033	5361	4780	4268	3809	3395	3016	59

Using the Proportional Logarithms

If we go back to the beginning and now look to the 6th step we find we have 10hrs 24min with adding the True Local Mean Time and the Equivalent Greenwich Mean Time. With this information we can use above Logarithms.

Finding the 10hr column and then the 24min row we see the junction shows .3632 with this we now apply it to the individual planets movement.

We need to work out the difference between each of the planets on the 7th to the 8th. This calculation is below

Sun on 7th was 15° Scorpio 11′
on 8th =16° Sc 11′ exactly 60' = 1°
The total movement for 24hrs.
The following is the calculation for finding all planets movements
Sun 15° Sc 11′. First each 1° contains 60 minutes so we deduct 11 from 60.

60 min
11 -
= 49'

16° Sc 11′ +11
=60 min = 1°

Going back to the Logarithms we see 1° is 0' is 1.3802 we then add the Constant Logarithm .3632

1.3802
+.3632

1.7434 we go back to the logarithms which shows 0hrs 26m

Sun position on 7th 15° Sc 11′
+ 0° 26′
=15° Sc 37′ Sun position for 10hr 24m

The next example is the Moon this has the fastest calculation as it moves around the Earth not the Sun and the least distance from the Earth. Going by the ephemeris the moon is 0° Sagittarius 50′ on the 7th November. It had moved to 12° 39′ in Sag the next day, now we calculate the exact difference between the 7th and 8th of November:-

Moon on 7th 0° Sag 50′

Moon on 8th 12° Sag 39 difference 11° 49

7th 0° Sag 50′
\+ 10′ → 10′
1° Sag 00′ VV+11 00′
\+ 11 Sag 00′ + 39
12 00 = 11 49
forwarded 12 00′
\+ 39′
movement in 24hr
8th Nov 12° 39′

In logarithms following the 11° column to 49th row we find .3077 (log. For ☽)
Now we add the Constant Logarithm to the difference between 7th and 8th Nov to find the adjustment of the moon.

.3077
\+ .3632
.6709

Going back to the Logarithm we find .6709 is in the 5° column and 07 minute row. Moon 0° 50′ Sag plus 5° 07'
+0 50′ -= 5° Sag 57′ Moon's position for 10hr 24

Venus position on 7th November was 9 deg Libra 10 min Venus is inside Earth's orbit as such will move faster than the Earth remembering the Earth moves around the Sun 1 deg every 24 hours, we therefore expect Venus to move further as shown below.

7th Nov = 9° Lib 10′
8th Nov = 10° Lib 22′ difference 1° 12′

7th 9° Lib 10′
\+ 50′ → 50′
10 Lib 00 +22
\+ 22 72′ = 1° 12′
8th 10 Lib 22

Log for Venus = 1.3010
\+ .3632 (Constant Log)
1.6642

1.6642 reconverted through Logarithm is 32′
Venus on 7th 9° Lib 10′
\+ 32

9° Lib 42′ Venus position for 10h 24m

Uranus position on 7th Nov `1972 it's the 7th planet from the Sun therefore with all outer planets with reference to the Earth will move slower, as the Earth moves 1 deg in 24 hours we would expect it to move very little, where the Earth takes a year to circle the Sun, Uranus takes 7 years roughly whereas Pluto takes between 12 and 32 years.

Uranus on 7th Nov '72 20° Lib 20′ = Difference 4′
8th 20° Lib 24′

7th 20° Lib 20′
8th + 04′
20° Lib 24′   (4 minutes total movement in 24 hours)

Using the Constant Logarithm .3632 we add it to the log for Uranus

2.5536 Log for Uranus
+ .3632 (C.L,)
2.9195 this total reconverted 2′

Uranus on 7th = 20° Lib 20′ ←
+ 02′
20° Lib 22

Chapter 13 - Foods of Constellations etc.

In this section I have arranged foods as to Constellation, as each Constellation and it must be remembered that each Planet rules a Constellation and the food will apply to it. Along with Houses.

In creating a cure a combination found in plants can cut down the ingredients where you see the foods contain multiple ingredients; but, in fear of repeating myself, remember the opposite side & sign in a chart, if there are planets whether transiting, progressive or natal maybe affected negatively, and so there would be a degree of that planets /constellations and houses that plant would require. This is paramount where there is little orb from corresponding degree. Personally I've found the best formula is by using all the influences and a separate plant for the planet/constellation and house.

To combine it you need to look to the squares, oppositions, quincunx's.

Example: Almonds contain two specific ingredients Magnesium and Silica these are under two separate planets Sun & Jupiter; If you go to 'The Anatomy' you will see which part of the body these planets rule.

* A clearer illustration of Chemical compositions can be found in Constellations & Their Chemical Composition

Foods of Constellations

Foods - Herbs	♈	♉	♊	♋	♌	♍	♎	♏	♐	♑	♒	♓
Vegetables & Fruit												
A. Acacia			●									
Aconite										●		
Adder's Tongue				●								
Agrimony									●			
Almonds					●				●			
Aloes -	●		●			●			●			
Flowers											●	
Angelica					●							
Anise			●			●			●			
Aniseed			●			●			●			
Apple		●					●		●			
Apricot									●			
Archangel		●					●					
Artichoke		●					●					
Asparagus	●	●					●		●			
Avocado			●	●		●						
Avocardo Peel								●○				
Azaleas			●									
B. Balm			●			●			●			
Banana				●								
Banksia											●	
Barley										●		
Barberry	●											

Herbs/Veg &n Fruits	♈	♉	♊	♋	♌	♍	♎	♏	♐	♑	♒	♓
Barren Wort										●		
Basil								●○				
Bay Leaf					●							
Bay Tree					●							
Beach Tree										●		
Beans						●	●					
Beet (White)									●	●		
Beetroot									●	●		
Birch							●					
Birdsfoot										●		
Bitter Sweet			●									
Black Hellebore										●		
Blackberry		●					●					
Blackberry Spikes												
Blackthorn										●		
Bloodwort									●			
Borage									●			
Bramble							●					
Briony	●											
C. Cabbage				●								
Cacti	●							●	●			
Calendula					●							
Capers											●	
Capsicum	●			●								
Carraway			●									

Herbs/Veg &n Fruits	♈	♉	♊	♋	♌	♍	♎	♏	♐	♑	♒	♓
Carrot	●		●		●	●						●
Cayenne Pepper	●											
Celery			●	●				●○		●	●	
Centuary -	●				●							
American												
Cherries - Winter		●	●				●					
Chestnut		●	●									
Chickpeas			●									
Chickweed				●								
Cinnamon									●			●
Cloves	●	●							●			
Cocaine - Coca											●	●
Coffee												●
Columbine						●						
Comfrey			●	●						●		
Coral-wort				●								
Corn Hornwort					●							
Cornflower				●								
Couch grass		●										
Cowslip		●					●					
Cranebill		●					●					
Cresses	●			●								
Croton Tiglium -											●	
G. Ginger	●											
Gooseberries		●	●									

Herbs, Veg & Fruit	♈	♉	♊	♋	♌	♍	♎	♏	♐	♑	♒	♓
Wild lemons	●				●							
Lettuce				●					●			
Licorice			●									
Lily - of the Valley		●	●☛	●								
White												
Liverwort		●	●									
M. Madder	●											
Maidenhair			●									
Mandrake			●☛							●		
Mango				●	●							
Maple									●			
Marigold					●							
Marshmallows		●										
Mercury			●									
Mint							●		●			
Moss										●		●
Motherwort								●				
Mulberry			●				●					
Mushroom + Salt				●								
Salt											●	
Mustard	●								●			
Myrrh									●		●	
Myrtle (Wax)	●		●									
N. Nailwort			●									
Navelwort										●		

Herbs, Veg & Fruit	♈	♉	♊	♋	♌	♍	♎	♏	♐	♑	♒	♓
Gourd				●								
Grapes		●		●	●							
H. Hare's Foot			●									
Hart's Tongue									●			
Heart Trefoil					●							
Hemlock										●		
Hemp										●		●
Holly										●		
Honeysuckle	●	●						●○				
Hope	●											
Hops	●											
Horehound			●☛									
Horseradish	●			●								
Horse-Tongue	●		●									
House Leek									●			
Hyssop				●								
I. Iris				●								
Ivy										●		
J. Jasmine									●			●
K. Kidney Beans		●					●					
L. Laurel					●							
Lavender		●	●									
Leeks	●			●	●							
Lecithin	●			●	●							
Lemons -												

Herbs, Veg & Fruit	♈	♉	♊	♋	♌	♍	♎	♏	♐	♑	♒	♓
Nettle	●											
Nightshade								●				
Nutmeg									●			
O. Oaks									●			
Oats	●		●						●	●		
Oils * pain relief										●		
Olive					●				●			
Olive Spurger			●									
Onions	●		●							●		●
Orange					●							
Orris Root				●								
P. Parsley			●									
Parsnip			●☛							●		
Peanuts							●					
Pennyroyal							●					
Peony					●			●				
Peppermint	●				●				●			
Peppers	●											
W. Walnut					●							
Watercress	●											●
Whitlow Grass				●								
Wild Wallflower				●								
Willow				●								
Winter Greens				●								
Witch-hazel	●											

Herbs, Veg & Fruit	♈	♉	♊	♋	♌	♍	♎	♏	♐	♑	♒	♓
Wolf's bane										●		
Wormwood	●							●○				
Y. Yarrow							●					
Yew Tree										●		
Yacca											●	

ANIMAL MEAT &												
SEAFOOD	♈	♉	♊	♋	♌	♍	♎	♏	♐	♑	♒	♓
Ants -												
Fire	●							●				
Honeypot	●	●										
Beef -		●										
Brains		●	●									
Breast		●		●								
Heart		●			●							
Kidney		●					●					
Liver		●					●		●			
Mammary Gland		●		●								
Milk		●		●								
Neck		●										
Buffalo - Feral												
Water	●										●	●
Chicken -			●									
Anus			●					●				

ANIMAL MEAT & SEAFOOD		♉	♊	♋	♌	♍	♎	♏	♐	♑	♒	♓
Breast			●	●								
Heart	♈		●		●							
Livers			●						●			
Lungs			●									
Thighs			●						●			
Wings			●									
Banter eggs				●								
Crab				●								
Blue Claw Crab				●							●	
Blue Swimmer				●							●	
Mud Crab				●						●		
Red Crab				●								
Sand Crab				●		●		●				
Fish -	●										●	●
Mullet			●			●					●	●
Red Fish											●	●
Blue Fish											●	●
Silver Fish	●		●								●	●
Sword Fish												●
Shark								●				●
Goat -	●									●		
Ankles	●									●	●	
Brain			●							●		
Heart					●					●		

ANIMAL MEAT & SEAFOOD		♉	♊	♋	♌	♍	♎	♏	♐	♑	♒	♓
Goat												
Breast			●	●								
Heart	♈		●		●							
Knees										●		
Livers			●						●			
Lungs			●									
Milk										●		
Thighs			●						●			
Wings			●									
Banter eggs				●								
Crab				●								
Blue Claw Crab				●							●	
Blue Swimmer				●							●	
Mud Crab				●						●		
Red Crab				●								
Sand Crab				●		●		●				
Fish -	●										●	●
Mullet			●			●					●	●
Red Fish											●	●
Blue Fish											●	●
Silver Fish	●		●								●	●
Sword Fish												●
Shark								●				●
Lobster - cooked		●										●
Octopus			●					●				
Oysters	●											
Rock Oysters											●	

◌in the Scorpio column means the product needs to be Fermented. I felt the symbol of the ¼ Moon appropriate.

☞ Finger symbol in a column means there can be dangers to areas of the ruling planet and should be used sparingly

Where it involves meat and different parts this would work best with the Sun Sign of the individual e.g. Taurus person with a tumor in the brain may look to the brain of the cow/beef.

The grey areas are those foods I've used, and have been nearly all the meat and crustaceous foods except goat and other exotic foods that aren't found in the local supermarket. I've also trialled the Scorpion. Not good for hair when Scorpio's 8th house and Jupiter are in Aries which are opposed to Saturn, Jupiter rules hair and 8th house is ruled by Pluto/Scorpio everything or nothing Aries rules the head and Saturn shows blockages when opposite.

Some of the more interesting foods are from the Tables of Composition of Australian Aboriginal Foods by Janet Brad Miller, Keith W. James & Patricia M.A. Maggiore

The best foods I've found for cures have been those that are known as the Symbols given for each zodiac sign. E.g. Aries symbol is the Ram, the Ram is the male sheep, but, secondly the term is also used for a male goat, these two shouldn't be confused the Ram in astrology is that of Aries and the sheep family, the Ram in Capricorn is the Goat. As such, I often use sheep meat, lamb (common in supermarkets) for formulas; alternately Aries is a forger and rules sharp instruments it also rules items that appear sharp, this can be used in foods i.e. Wheat under Mars. These are alkaline as are all the uneven signs of the zodiac, but, as with many plants it can be altered this is seen with Tobacco it has a Mars and Neptune influence, which means it could give a neutral ph affect.

The next is the Taurus symbol which is the Bull, I have found beef meat has been successful in throat problems but mixed with oil; where it concerns a plant I look for sweet and earth plants, another animal that could be looked at is the snail, it eats vegetation and is slow as the bull in general, and so the snail can be used for throat problems and areas associated with Taurus and could also be ruled by Virgo as it's small. Their lactation is a substance that could be used in the pituitary gland or for excessive nose running, possibly just rubbed on the nose. I'm not suggesting you try it only that scientists could look into it, it is already an animal eaten by the French.

When doing formulas, a person needs to look at the planets affecting each other in particular the 150° angles, but this can also be found by planets that rules constellations that would otherwise be in 150° angle in a house that is making 150° angle. Therefore, you're looking for at least two types of foods.

GLOSSARY

(✱ This symbol the asterisk shows the aspects that are hardly used)

Ascendant – A degree and constellation calculated at the time of birth which is crossing the eastern hemisphere, illustrating the 1st house, it shows how you look and what your immediate environment would be like. And, is the where all other cusps are calculated.

* Bi-Quintile – a minor aspect needing more exploration, generally no longer used.

Conjuncts – Planets that are at the same degree or within a 8 degree orb for planets up to Saturn. Conjunctions are a focal point in the Birth Chart. Their effect is to lay positive or negative stress according to the Sign and House positions of the planets concerned.

Cusp – is the precise degree found from the Time of birth, and is organised by calculating the Ascendant. It may be confusing but each cusp occupies thirty degrees, important cusps are 1st, 4th, 7th and 10th, these are the active houses or non-submissive houses, that look at the immediate environment, home, long term partnerships and career path. It also pinpoints north, south, east and west degrees in a chart.

* Decile – a minor aspect needing more exploration, generally no longer used.

Geocentric Chart – places the Earth in the centre of the solar system.

Heliocentric Chart – places the Sun at the centre of the solar system.

Horoscope – a Greek term meaning 'hour watcher' a chart formed on the D.O.B. and time of birth is then created placing planets as seen in their constellations.

Intercepted signs – intercepted signs are more common in charts where the birthplace is found in more extreme north and south latitudes versus latitudes found near the equator. In the extreme latitudes a house will contain more than 30 degrees, and in this case we find an entire sign enclosed in it and therefore, without appearing on any actual cusp. Such a sign is called "intercepted". In the equal house system, intercepted signs in a chart cannot happen.

Meteorological Astrology – the application of the science to the forecasting of weather conditions, earthquakes and severe storms.

Midheaven – is the M.C. Medium Coeli it is the highest point in the sky taken from the Earths tilt at the time of birth

Mundane Astrology – is Astrology which is not based on understanding personalities which Horary is, it's based on the time something happens in the outer world i.e. a letter being received, a ship sinking, a volcano erupting etc.

North Node / South Node - These points are always opposite each other. They are mathematical points where the Moon's orbit intersects the plane of the ecliptic. Some say they are points of karmic like yin and yang. The North Node shows the areas that we need to develop, whilst the South Node are areas we need to develop less in, and have to learn not to fall back on. They are points that take into account the Sun, Moon and Earth at the time of birth.

Oppositions - accentuate the polarities in a Chart, and can also form a strong integrating link. A person needs to come to terms with the opposition by combining both.

Planets – The planets in astrology are those that stay within a 10-19 degree orb and are seen along the Sun's ecliptic/path, the same path that pinpoints the 12 constellations. The Sun is not actually a planet it is a Star, it is halfway thru its life cycle and is believed to have approx. 5 billion years left, but is considered a planet for convenience, in fact it is the opposite of Earth along, the Moon is also considered a planet which is in fact the Earths Satellite. As such there are 10 planets, 11 if we count the newly found Chiron Planet it is believed to be between Saturn and Uranus.

Progressions – One day 24 hrs from the birth time, equals one year of a person's life and equals the first year's progression, and, illustrates how the personality changes. It also indicates what the parents were doing and illustrates stages in their relationship.

Quincunx – is 150 degree angle with a 2-3 degree orb, it is said they are unpredictable but it's more likely that they are set off by transits or progressions. They can cause health problems.

* Quintile – minor aspect needing more exploration, generally no longer used.
*
Radix – a point of origin for a personal relationship.

* Semi-Docile – minor aspect needing more exploration, generally no longer used.

* Semi-Sextile – is a 30 degree angle, it is a minor aspect. This is a favourable aspect and enhances the meaning of whichever planets are involved. Semi Sextiles are the reverse to Semi-Squares they can make a person feel happy or optimistic but, not necessarily knowing why unless you look at the chart to see what planets are involved.

* Semi-Squares are weak aspects they can cause a person to feel annoyed but not sure why.

* Sesquiquadrate – is 135 degree aspect with 2 degree orb, it indicates strain and needs to be examined especially if found with more then 1 close planet.

* Sextile – is a 60 degree angle and makes an easy flow of air in a 360 degree circle, because of this angle it doesn't look into things deeply. Like Trines it can bring out the pleasant characteristics in a suitable subject. People can find it easy to work in the areas, which the planets show.

* Square – is a 90 degree angle depending on position, can indicate tension or give drive and strength to the character to overcome.

Solar Arc - Distance in degrees between the progressed Sun and the natal Sun. By adding this distance in degrees to the rest of the planets and points. Because the Sun moves at a rate of almost one degree a day, this method makes it rather easy to estimate positions of progressed planets and points. The degrees of progression are roughly equivalent to the age of the individual.
Square – is a 90 degree angle which can indicate tension, the person would have to try and bring these two aspects together to overcome the negatives and extremes.

Trine – A 120 degree an Angle, that shows favourable aspects. They are also helpful to a strong character but may exercise a spoiling effect on shallow people

Vertex – represents the intersection of the ecliptic and the prime vertical. It's considered an auxiliary Descendant. And found in the western hemisphere *as charts are on the reverse it would be found on the right hand side of a chart. As in the English language we read from the left. The Anti-Vertex is the point that is opposite the Vertex. Astrologers have found it important in relationships either meeting the significant other or the ending of that relationship. Some have found it also can indicate birth or death.

Often asked questions: -
Love – The question of true love and how to find it is often asked, below shows varying aspects that can cause a person to love whilst not necessarily being loved in return for the same reasons. Anything to last needs to be in a fixed sign.

Love is found when 1. A's Sun Sign is the same as B's Ascendant 2. When A's Sun Sign is on the Cusp of the 7th house in B's Chart. 3. If the girls Sun is in the same as his M.C. she will give encouragement and force to his aims. 4. Sun and Ascendant same sign. 5. Sun and Ascendant opposite gives considerable rapport. 6. Sun and Ascendant Trine or Sextile. 7. Aries and Scorpio quincunx but can be good. 8. Pisces and Leo quincunx, can be good. Usually quincunxes are not good.

Honour – is found when there are a number of planets in the 10th House

About the Author

As the author I wouldn't have felt compelled to write this book if I hadn't personally lived by what I've written. It's my hope that my research may help Astrologers become more involved in medical research. Everything has to begin with a basic formula which can be proved over and over again.

My compulsion for research first appeared when my family were very young where I began to look into personalities using Astrology to solve in-house fighting way back in the late 70's. In the 80's and 90's I ran a local newspaper, where without the co-operation of my family it would never have got out. Many a night we were walking around delivering it. The paper actually caused me to be involved in many local and rural areas and to look into the politics of our country with media releases from many branches of government. The volunteer programs, in neighbourhoods we covered, always needed prop up by the government who were always calling poor, whilst increasing their wages 3 fold.

Prior to closing the paper, I did a 2-year volunteer radio program researching every current affair program using astrology and its planets I realized how diverse the subject really was for even medical research on sudden curers were already defined as curers in astrology if they were only recognised this book is about recognising Astrology as a driving force for recognising those curers and delving deeper into complex curers.

When I saw my mother die in the early 90's having been turned away from Naturopaths where an operation was all too late. And seeing family members die before their time. I thought if we could read a person's personality using astrology why can't we look into medicine which history showed it was once used for.

I've put in a section under Foods ruled by Planets etc, in grey are the plants I've personally used. Of course I can't make the same formula for myself that I'd make for someone else and as I've got older and put myself in more confined spaces I've found it more difficult to do formulas for others let alone myself.

People want to know if there are quick curers, in some cases its hit and miss. You need to compare colours with foods and planets and get the final colour that would compliment both the disease symptoms and combined planet.

You need to know a bit about the anatomy for instance the Livers a blood red/brown. So a good food for people with Liver disease is to buy and eat liver or beetroot the same colour and preferably the same shape. I've tried to give all the elements that can be used e.g. minerals in cell salt, vegetables & herbs and where it can be found in animals, but why use an animal when we can use a vegetable or mineral.

I have diplomas and did a basic course in Astrology but I've been self-taught in the medical field mainly because there has been no one teaching medical astrology; But, I should add the extended studies have advanced through following the Runes, over many years, where I've been able to find the correlation; and, where I saw the correlation in the universe to the whole range of events good or bad which I was able to relate back in the terms of astrology and the many Pisces traits I have.
I wish all those who try making formulas the very best don't forget you might need the runes as well
The reason for my research, I guess you'd have to say is Death and what's caused me to write this book is that whilst we have to accept death it shouldn't be premature

References

(If you haven't been to university you mightn't realise that the superscript numbering indicates the original authors or program as known at the time)

[1] Casenotes of a Medical Astrologer by Margaret Millard M.D.

[2] Encyclopaedia of Medical Astrology by H.L. Cornell, M.D. 1932

[3] The Australian Women's Weekly – Astrology Book by Richard Sterling

[4] The New Compleat Astrologer by Derek and Pauline Baker

[5] Collier's Encyclopaedia 1962 updated 1986

[6] The World Encyclopaedia 1971

[7] Solar Fire Astrology Software

[8] Chart our Own Horoscope by Ursula Lewis

[9] The Vitamin Bible by Earl Mindell

[10] Tables of Composition of Australian Aboriginal Foods by Janet Brad Miller, Keith W. James & Patricia M.A. Maggiore

www.ingramcontent.com/pod-product-compliance
Lightning Source LLC
Chambersburg PA
CBHW050715180526
45159CB00003B/1038
* 9 7 8 1 5 3 3 5 7 3 7 5 9 *